Archaeological Conservation Using Polymers

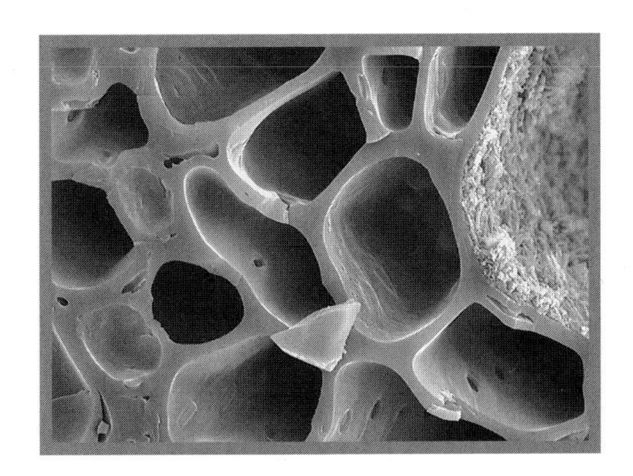

NUMBER SIX
Texas A&M University Anthropology Series
D. Gentry Steele, General Editor

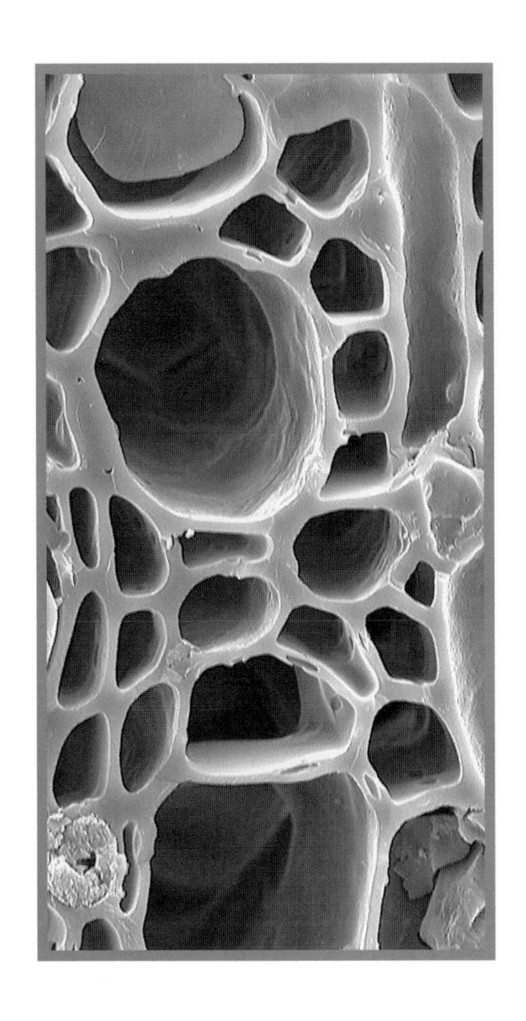

Archaeological Conservation Using Polymers

Practical Applications for Organic Artifact Stabilization

C. WAYNE SMITH

Foreword by J. M. Klosowski

TEXAS A&M UNIVERSITY PRESS • COLLEGE STATION

First edition

The paper used in this book
meets the minimum requirements
of the American National Standard
for Permanence of Paper for
Printed Library Materials, z39.48-1984.
Binding materials have been
chosen for durability.

Library of Congress Cataloging-in-Publication Data

Smith, C. Wayne.
Archaeological conservation using polymers : practical applicatons for organic artifact stabilization / C. Wayne Smith.—1st ed.
p. cm.—(Texas A&M University anthropology series ; no. 6)
Includes index.
ISBN 1-58544-217-8 (alk. paper)—
ISBN 1-58544-218-6 (pbk. : alk. paper)
1. Antiquities—Collecton and preservation. 2. Archaeology—Methodology. 3. Archaeological chemistry. 4. Polymers. I. Title. II. Series.
CC135.S56 2003
930.1'028—dc21 2002006420

The author and publisher gratefully acknowledge a generous grant from Bill Caruth in support of publication of this volume.

To Helen,

Thanks for sharing this

and all my adventures.

Contents

Foreword

Preservation with reactive chemicals can be called by many names but the most prominent is von Hagen's term, plastination. Other terms are chemical bulking and reactive filling. They are all used to describe the process of coating and impregnating cells with materials that react with the walls to prevent water from attaching and reacting. In many cases, we are also talking about filling the cell or opening to prevent collapse.

The goal is to place a substance on the cell that reacts to the carbonols (—COH) present on much of the cell surface. Reactive silanes (cross-linkers) are particularly adept at this and can be combined with polymers (carbonol or silanol [—SiOH] ended polymers) that also react with the silane cross-linkers. Thus we have silane cross-linkers that react to the cell wall, to each other, and to the polymer. Almost always, this requires one or more catalysts to hasten and complete the reactions.

We can use very reactive cross-linkers like methyltriacetoxysilane $MeSi(OCOCH_3)_3$ or less reactive cross-linkers like methyltrimethoxysilane $MeSi(OCH_3)_3$. Typically, when deep penetration into cell walls or body cavities is desired, slower-to-react materials are used to postpone the reaction until penetration is fairly complete. Alkoxy functional (alcohol releasing) cross-linkers are preferred, as their reaction to the cell wall or to the polymer usually requires a catalyst. The von Hagen procedure uses a polymer and a catalyst to impregnate and coat the cells followed by the introduction of the cross-linker. Our technique involves using polymers and cross-linkers to penetrate cell walls and then adding catalyst. Each technique has its advantages.

Both techniques often require specimen preparation, or preservation, which might require rinsing them clean and setting the tissues with formalin solution. Next is the medium exchange, where water in cells and cavities is exchanged for a more compatible and less reactive medium, usually acetone, but sometimes alcohols. The exchange is continued until most of the water is eliminated.

The next step is penetration with reactive materials. The classic von Hagen procedure uses a mixture of a catalyst (typically dibutyltindilaurate) and a silanol-ended polymer. This technique involves immersing the acetone-saturated specimen in a bath of this mixture under a very slight vacuum and boiling off the acetone with the vacuum to leave little voids that are then filled with polymer/catalyst mixture. When the acetone is gone, the cavities are drained of excess solution and a cross- linker, typically tetraethoxysilane $Si(OEt)_4$, is added. As this penetrates the cavities, the polymer, cross-linker, and catalyst converge and react. The reaction is permitted to continue until it is complete (typically in some heated chamber since often the cross-linker is put into the system from a vapor).

Our technique uses formalin (if desired) and acetone to displace water. A mixture of polymer and cross-linker is then added to accomplish cell penetration. The polymer and cross-linker mixture is preferable because it is more stable than the polymer and catalyst mixture. Our mixture thus has a longer shelf life, and material that is left over or drained from the cavities can be stored and reused. The lack of reactivity of the first two ingredients allows most of the work to be done at ambient temperature and not in the freezer, providing both an equipment and a time advantage.

The next step is the addition of the third ingredient, the catalyst. Dibutyltindiacetate is preferred to dilaureate because the diacetate, being smaller, penetrates cells faster. Typically, catalyst is painted on the surface and rinsed through open cavities, putting a maximum catalyst dose at the surfaces to enhance penetration. The specimen is then put out in the air to cure. The cure in the von Hagen procedure takes place when the cross-linker penetrates the system. The von Hagen procedure could also use the painting or dipping technique to expose the specimen to this third ingredient (the cross-linker), but that might require a posttreatment procedure of exposing the specimen to air. Our procedure exposes the specimen to air to bring it into contact with the water in the air to allow excess cross-linker to react. Our method uses an excess of cross-linker whereas von Hagen's does not; the water in the air completes the reaction. Our mostly room temperature procedure adds cross-linkers to polymers in the initial step and postpones the use of the smaller catalyst to the final step. The procedure is relatively quick. The specimens can usually be allowed to cure in the air after one or two weeks.

The following equations illustrate the cure reactions:

1. The reaction to the cell wall is SiX + HOC ➔ SiOC — + HX X = alkoxy

2. The reaction to the polymers is SiX + HOSi ➔ SiOSi — + HX

3. The reaction with an excess of cross-linker and insufficient water is SiX + SiX ø no reaction

4. Air exposure yields partial hydrolysis followed by condensation:

a. SiX + HOH ➔ SiOH + HX

b. SiX + SiOH ➔ SiOSi + HX

The work done by C. Wayne Smith, others, and me concerns the investigation of different combinations of cross-linkers, polymers, and catalysts, along with variations in the exposure techniques. For instance, in most of Dr. Smith's work we use methyltrimethoxysilane $MeSi(OMe)_3$ as the cross-linker instead of the tetramethoxysilane $Si(OEt)_4$ favored by von Hagen. The $MeSi(OMe)_3$ is only trifunctional while the ethylorthosilicate is tetrafunctional: using $MeSi(OMe)_3$ thus yields a lesser cross-link density and a more pliable specimen. The remainder of the procedure is like cooking: ingredients are mixed and matched to achieve a desired effect. To get a more pliable, leathery touch, use a trifunctional cross-linker and a higher molecular weight polymer. To get a less leathery, stiffer touch, several procedures can be performed: add more cross-linker to the polymer/cross-linker mix to get some in-situ formed resin in the system; use a tetrafunctional cross-linker; use a lower molecular weight polymer; or use some combination of these procedures.

In addition to the variations in the functionality of the cross-linker and length of the polymer chain, there are also variations in the catalysts that can be used. The workhorse catalyst for our purposes is dibutyltindiacetate. However, this is a forever active catalyst, and the system is forever equilibrating; this also happens with the dibutyltindilaurate generally used with the von Hagen procedure. Divalent tin catalysts yield a nonequilibrating system because they eventually become noncatalytic tin oxide, but they are difficult to use because they hydrolyze and lose activity if excess water is present or if the catalyst is prematurely introduced. They are best reserved for special situations.

Titanate catalysts are another possibility, but again their high reactivity makes for difficulty in use, so they are also best reserved for special situations. The polymers, cross-linkers, and catalysts used with silicone sealants are the types of materials that can be used in preservation. However, in preservation, we can and do sometimes use no polymer, just catalysts and cross-linker. It is an especially effective way to preserve old glass retrieved from the sea. Only short polymers can be used with glass systems since a leathery touch is undesirable in this application. This is all part of the science of preservation with reactive chemicals. With glass, polymeric cross-linkers are also quite useful. With hair, and hide with hair, one often wants to rinse out the preservation chemicals before the cure proceeds. This is again part of the art of preservation with reactive chemicals.

There is an art and a science to preservation with reactive chemicals, both of which must be practiced to achieve a desired result. Keep this in mind as you read through this book. Science provides proper combinations to give good penetration of cells and cavities and to sufficiently complete reactions to prevent degradation. Science also provides some order of additions that favor penetration and the various combinations to give the toughness when needed. Art is needed to choose acceptable combinations to give the desired effects to all manner of things one might wish to preserve, such as skin, hair, glass, wood, paper, or leather.

Preservation with reactive chemicals is like cooking. The science of cooking provides nutritious combinations to sustain health, but the art of cooking provides the combinations that taste good. The science of preservation yields many combinations that will penetrate and preserve, but artistry determines which combination gives the result desired in a specific specimen. That art is what is primarily depicted in this book.

J. M. Klosowski
Independent Consultant
and Scientist Emeritus for the
Dow Corning Corporation

Acknowledgments

As a graduate student, I had the great privilege of working with Donny Hamilton at the Archaeological Preservation Research Laboratory at Texas A&M University. In the field and in the laboratory, I was given the opportunity to excavate and conserve a wide range of artifacts from nautical excavations at Port Royal, Jamaica. It did not take long to see that, to some degree, our conservation efforts were in vain. Since there were no facilities for long-term controlled curation, many artifacts continued to deteriorate even after appropriate conservation strategies had been applied.

The long-held tenet that all conservation methods had to be reversible also proved troubling to me. As a graduate student, I had observed firsthand that, in many situations, attempts at first removing bulking agents from fragile organic artifacts and retreatment using new conservation methods and materials resulted in additional damage to the artifacts that I could not justify. I soon found that conservators from many countries held similar doubts about the true reversibility of conventional treatment strategies. The most important early lesson learned was that there were very few conventions in the conservation of organic artifacts; each artifact presented a unique challenge. It became apparent that if we were to truly address the conservation edict of long-term well-being of the artifact, different treatment methods and materials had to be devised to ensure long-term survival in a noncurated environment.

Lack of funding was also an issue. Archaeologists everywhere said they could not afford to conserve all the artifacts recovered from their excavations. Apart from the tragic loss of artifacts, this is irresponsible archaeology. Delving into their comments more deeply, however, I was not surprised to learn that lack of funding was not the sole reason for the backlog of artifacts waiting to be conserved. Lack of knowledge and lack of faith in existing technologies were key reasons expressed for holding off on the preservation of some artifacts.

All my mentors have taught me the same valuable lesson: knowledge is not static. Archaeological chemistry and the development of better conservation solutions have been a serendipitous adventure, resulting in the awarding of three patents—the first ever awarded in the Department of Anthropology. While Donny Hamilton, head of the Nautical Archaeology Program at Texas A&M University, and Jerome Klosowski, senior scientist at Dow Corning Corporation, are coinventors of the Passivation Polymer technologies, their contributions have been more than scientific. I thank them both for their encouragement, scientific visions, and deep-seated belief in this research. I would also like to thank Ole Crumlin-Pedersen, Senior Researcher at the Center for Maritime Archaeology in Roskilde, Denmark, for bringing the problems of long-term stability of PEG-treated artifacts to my attention. His insights have had a tremendous impact on my research and on the science of maritime archaeology. Many thanks, also, to Research Coordinator Mark Gilberg of the National Center for Preservation Technology and Training, for supporting initial research in deveoping Passivation Polymer technologies.

The contributions of the extended research

team at Dow Corning Corporation have been innumerable. I thank Leon Crossman and Arthur Rathjen for championing the cause of advancing archaeological chemistry and joint research with Texas A&M University. Their generosity and commitment to providing educational resources for schools and universities are inspirational.

Without the financial support and encouragement of Woodrow Jones and Ben Crouch, both of the College of Liberal Arts, development of Passivation Polymer technologies would not have been possible. Their vision for advancing archaeological chemistry research within the College of Liberal Arts forms the unique alchemy of art and chemistry that is the basis of my research.

It is impossible to personally acknowledge all the conservators who have contributed to my research. Your questions, concerns, and requests for assistance in the preservation of artifacts have been the driving force behind this research. I hope this book answers some of your questions. My greater hope, of course, is that this work serves to generate new questions and directions in conservation research. Where new insights abound, there is no stasis.

C. Wayne Smith
Texas A&M University

Archaeological Conservation Using Polymers

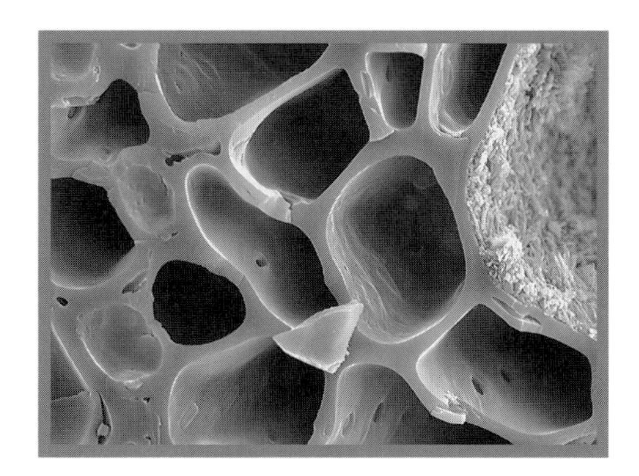

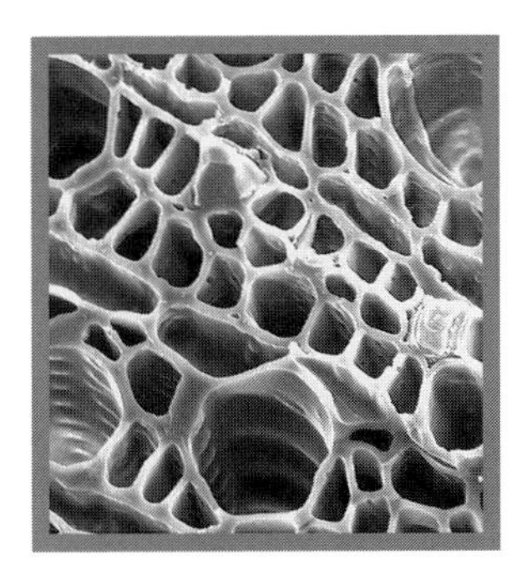

Introduction

The archaeological conservation of highly perishable organic matter has always been problematic. In 1996, for example, the Texas Historical Commission began excavating René-Robert Cavelier, Sieur de La Salle's vessel, *La Belle,* which sank off the Texas coast in 1686. The recovery of this vessel has presented a host of unique challenges in conserving wood, basketry, leather, and an assortment of supplies necessary to start a colony in the New World. Entombed in the mud that sealed it from decay for more than three centuries, the waterlogged hull and hundreds of thousands of fragile artifacts, including brain matter in the skull of one unfortunate sailor, would have been a futile conservation effort without new preservation technologies.

The aim of this book is to help conservators develop successful treatment strategies for organic materials using Passivation Polymers. These polymers are part of a series of chemistries and techniques developed at Texas A&M University to provide efficient and cost-effective conservation strategies for the purpose of archaeological preservation. Museums and historical societies have expressed the need for durable artifacts used in traveling exhibits and interactive displays. In many cases, traditionally treated artifacts cannot withstand the rigors of extended traveling and handling. Passivation Polymer–treated artifacts are more stable than their counterparts preserved using older technologies and can withstand sensible handling extremely well. These preservation processes are not a panacea for all artifacts, but the ease with which many of these techniques can be used makes them viable for various conservation challenges and for laboratories working with limited budgets.

In addition to discussing the advantages offered using Passivation Polymer technologies and materials, I consider a concept seldom addressed in conservation: artistry. Variance in equipment, relative humidity, laboratory layout, intended end results, and level of expertise are a few variables that play an important role in the development of preservation strategies and protocols that work for each technician. The key to consistent and aesthetically correct samples using Passivation Polymers, and indeed all conservation materials, is a willingness to explore treatment parameters and combinations of polymers.

In chapter 1 the Archaeological Preservation Research Laboratory at Texas A&M University is used as a model to illustrate an effective layout for day-to-day conservation of small organic artifacts using polymers. Too often, the entire working budget of a conservation facility is spent in the acquisition of elaborate equipment. This is a shame since well-trained conservators should be the first consideration of any laboratory. Accordingly, a short list of essential instrumentation is discussed in chapter 1. The distinction is made between equipment needed for conservation of artifacts and equipment and necessities for the purpose of research and development; small necessities and chemicals are also discussed. While conducting postdoctoral research at

Dow Corning Corporation, I learned from professionals how to work effectively and economically in the laboratory. Since it seems that all archaeological projects are underfunded and understaffed, these cost-saving suggestions can remove some of the economic stresses involved in outfitting a conservation facility.

Chapter 2 examines some of the mechanical processes used for Passivation Polymer processing of organic materials from marine and land sites. Two experiments are provided to demonstrate water/solvent and solvent/polymer exchange processes that are fundamental to many of the experiments and case studies throughout the book. In "Dowel Experiment," penetration depth of polymers and the presence of water in organic materials are discussed. The experiment illustrates the relationship between solvent dehydration and time in the treatment of wood. A second experiment, "Mass Spectrographic Analysis of Out-Gases Created from the Dehydration of Archaeological Wood Samples," illustrates the differences in exchange potential of organic artifacts that have been dehydrated using a single solvent dehydration process as compared to combined solvent processing.

Chapter 3 discusses the preservation of wooden artifacts using Passivation Polymers and is divided into two sections: waterlogged wood and dry-site artifacts. Bacteria, fungi, and the loss of structural integrity can be caused by waterlogging or desiccation from extreme dryness. Regardless of the object's archaeological provenance, stabilizing archaeological wood is a great challenge for the conservator. Mechanical destruction caused by freeze-thaw cycles in lakes, swamps, and bogs may act to accelerate the degradation of wood. Regardless of what type of degradation has occurred in a wooden artifact, Passivation Polymer technologies have proved effective in preserving diagnostic attributes.

Conservators face another challenge: museums around the world are filled with artifacts, big and small, that were preserved using polyethylene glycol (PEG) methods. Conserved during a time when PEG technology was new, these artifacts have not fared as well as artifacts preserved using the well- defined PEG treatment strategies developed in the latter half of the twentieth century by Per Hoffmann, David Grattan, and others. Re-treatment strategies developed to stabilize these difficult artifacts are discussed in a section entitled "Reprocessing and Stabilization of PEG-Treated Wood" and in the case study, "The Re-treatment of Two PEG-Treated Sabots."

New wood treatment strategies using self-condensing alkoxysilane polymers are covered in another case study, "Treatment of Waterlogged Wood Using Hydrolyzable, Multifunctional Alkoxysilane Polymers." These treatment strategies enable the conservator to successfully preserve waterlogged wood by introducing an alkoxysilane that, in the presence of atmospheric moisture, self-condenses to form polymers that accurately maintain the diagnostic attributes of waterlogged wood.

Numerous approaches and suggestions for the preservation of waterlogged and desiccated leather are outlined in chapter 4. Case studies are offered to demonstrate the adaptability of the baseline treatment process to accommodate even badly fragmented and desiccated artifacts. Important considerations for the storage and display of leather artifacts are also covered.

One area of great concern for all conservators is the treatment of composite artifacts. Often, it is necessary to disassemble an artifact into its component parts prior to treatment. However, this can damage fragile portions of the artifact. Additionally, structural changes in one or more components of the artifact often make the reassembly stages of conservation difficult. In chapter 5 the case study "Preservation of a Composite Artifact Containing Basketry and Iron Shot" details how wicker basketry was stabilized prior to removing heavy layers of concretion encasing the basket of shot. After the basket was processed in stages, the iron shot that had been removed from the basket were preserved using electrolytic reduction. The ease with which calcareous material was removed from the surfaces of the iron shot and the resultant treatment of the iron suggest that exposure to Passivation Polymers may have acted to preserve the iron, arresting surface exfoliation and pitting up to, and during, treatment.

Preservation of cordage and textiles from marine and land sites is discussed in chapter 6. Numerous conventional treatment strategies adapted for use with polymers are illustrated in a series of case studies.

It has been argued that glass is not an organic substance. While this is true, fluxes used in the making of ancient glass are organic materials, which accounts, in part, for the unique chemical structure of ancient glass. Treatment methods, accelerated weathering testing, and scanning electron microscopic analysis of polymer-treated glass are presented in chapter 7. Several case studies illustrate the abilities of alkoxysilane materials to stabilize badly deteriorated glass. Treatment techniques used for preserving extremely thin sections of onion bottle glass are discussed to illustrate the degree of creativity that can be employed in the treatment of friable objects. Like polyvinyl acetate and other traditional treatment strategies, reversibility of these treatment methods remains an issue. Data from accelerated weathering and nuclear magnetic resonance imaging experiments indicate that such glass remains remarkably stable much longer than glass treated using conventional methods.

A discussion of ongoing research in the preservation of badly deteriorated ivory and bone from marine and terrestrial sites is the focus of chapter 8. In one case study treatment strategies applied to the preservation of finely carved ivory from the Tantura Lagoon excavations in Israel are discussed. Another case study illustrates methods used to preserve sections of ivory tusk from a shipwreck off the coast of Australia.

In chapter 9 I explore some of the new tools and technologies that can help conservators devise more effective conservation strategies. Until recently, the conservation process has been hindered by the inability of traditional radiographs to supply sufficient data to give the conservator a complete view of an artifact concealed in a concretion. Through computerized tomography (CT) and computer-aided design (CAD), however, three-dimensional models of artifacts can be created using laser-driven polymerization of each "slice" of CT data. Stereolithography is another technology that is discussed.

Many conservation materials, considered part of the conservation tool kit, are polymers. Research has shown that these materials, once thought to be reversible, have a tendency to cross-link with the organic substrate of the artifacts they were intended to stabilize. While most of us tend to visualize chemists in white lab coats when we think of cross-linked polymers, the process of cross-linking is a natural process. As archaeological chemistry plays an ever-increasing role in our discipline, research scientists and conservators alike will see the expanding role of polymers in archaeological conservation. The polymer chemists who helped define the initial mechanisms at work in Passivation Polymer processes found the task of adapting their thinking to fit archaeological perspectives challenging. Though new terms will inevitably find their way into the conservation lexicon, I have tried to keep the use of new terminology to a minimum in this book.

Experience at the Archaeological Preservation Research Laboratory (APRL) at Texas A&M University has taught us that while complete reversibility of conservation treatments is desirable, it is a state that is sometimes easier philosophized than realized. Retreatability may be a more realistic end goal when determining suitable conservation methods for any given artifact.

The experiments and case studies offered in this book have proved successful for the preservation of a wide range of artifacts. I hope these materials give the reader a broad understanding of how Passivation Polymers, and other silicone oils, may be used for the preservation of organic artifacts. Because no two pieces of wood are exactly alike, selected treatment strategies must be adapted to treat artifacts on a case-by-case basis. Polymer-based materials offer the conservator infinite combinations of material, so that the preservation needs of individual artifacts can be addressed.

Anyone experienced in the complexities of field conservation would agree that artifact triage and stabilization play a major role in the eventual state of an artifact. Passivation Polymers allow the conservator to stabilize friable materials in the field so they can be more easily transported to the laboratory, where continued preservation can be undertaken. Polymers can help stabilize artifacts, even if more traditional conservation methods are used once the artifact is secured in the laboratory.

Invariably, success in using any conservation strategy requires patience and a marriage of technical and artistic know-how. Therefore, time should be spent experimenting with baseline processes before attempting conservation of artifacts. There is no substitute for hands-on experience, which allows the conservator to better understand how to use polymers in the preservation of organic material culture and how these treatments can be adapted to suit each artifact. This type of experimentation will also help the conservator understand and better utilize the equipment used for silicone oil preservation.

The goal of this research has been to fuel new dialogues in conservation and to encourage research along new avenues. Most important, this research encourages the reader to gain a new understanding of the material culture we strive so hard to preserve. The development of polymer chemistry for use in archaeological conservation is in its infancy. To date, development of Passivation Polymers has helped bring new perspectives to bear on the treatment of organic artifacts.

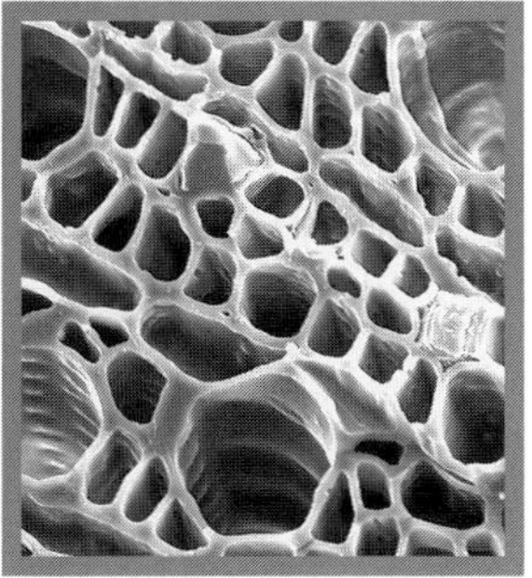

Laboratory Setup

Experimentation conducted on the development of Passivation Polymers required the use of sophisticated laboratory equipment. This book, however, offers some basic processes that can be implemented without the need for elaborate technology. While there is no one ideal setup for a functional lab, layout and location of equipment and materials should always adhere to guidelines established by reputable agencies such as the Occupational Safety and Health Administration (OSHA). Worldwide, a number of agencies act in the interest of promoting laboratory safety. Many of these have Web-based resources available, free of charge.

Although safety was a major consideration in choosing material for the development of Passivation Polymer technologies, many of the materials suggested for use in the laboratory require proper handling and storage. All technicians working in a chemistry lab need to be familiar with proper procedures for handling and disposing of chemicals. A complete inventory of chemicals should always be kept in case of an emergency, and material safety data sheets (MSDS) for chemicals used in the facility must be available for immediate consultation. The MSDS catalog contains vital contact information as well as important instructions if medical assistance is needed.

Figure 1.1 shows the floor plan of the Archaeological Preservation Research Laboratory at Texas A&M University. Much of our developmental research was conducted in this facility. Equipment such as the explosion-proof freezer system is essential for developmental research. For routine conservation work, such equipment is not necessary.

Major Instrumentation

The investment in elaborate equipment to outfit a laboratory for polymer research and artifact conservation is minimal. When outfitting a laboratory, safety should be the primary concern. High-flow ventilation and an eye-washing system are two of the most important considerations.

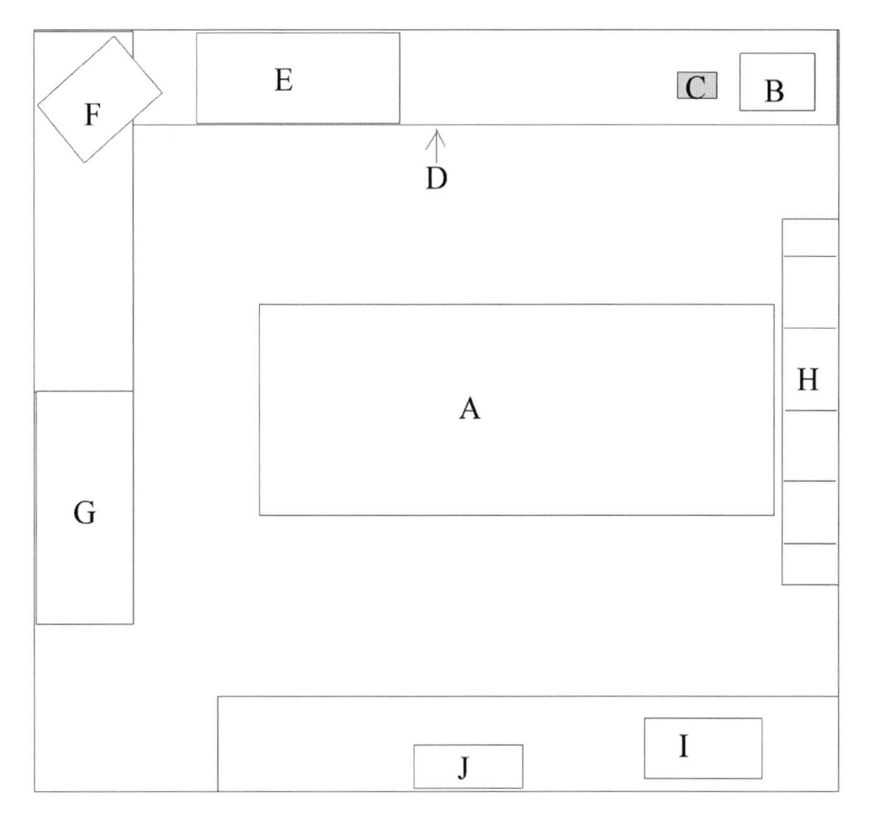

Fig. 1.1. A functional research–working laboratory: *(A)* double-sided workbench; *(B)* sink; *(C)* eye wash system; *(D)* under-cabinet flammable liquid storage; *(E)* vented fume hood; *(F)* vented warming oven; *(G)* explosion-proof freezer; *(H)* air-drying shelves; *(I)* computer with digital imaging camera; *(J)* electronic scales.

Because solvents are used in conjunction with many preservation processes, selecting equipment that will not break down when exposed to vapors is essential. Often, stainless steel beakers and small supplies can be found at a restaurant supply store for a fraction of the price of identical equipment sold through a chemical supply company. Generally, however, it is best to select equipment certified for laboratory use.

Vacuum Pump. Older piston-type pumps work best in the laboratory. Modern pumps are more efficient, but they are lightweight in construction, and many companies use plastic internal parts. These tend to break down rapidly in the presence of acetone or acetone vapors.

Vacuum Gauge. A simple manometer is essential, preferably one with an all-metal construction. In more inexpensive manometers, gradation marks are generally painted onto the back plate. During water/acetone exchange processes, any solvent fumes passing through the unit will cause the painted surfaces to crack and fade.

Dewar Flask and Gas Traps. The best way to trap harmful out-gases that contaminate the vacuum gauge and vacuum pump oil is to use a Dewar flask (fig. 1.2). These flasks are available through any major chemical supply company. A Dewar flask should be well insulated and large enough in volume to hold two cold traps hooked in parallel, along with a generous quantity of dry ice. With the Dewar flask and cold trap assembly placed in-line between the vacuum chamber and the vacuum pump, the cold environment should condense solvent fumes and prevent their escape into the manometer and vacuum pump oil.

Beaker and Vacuum Plate. A small, countertop vacuum unit can be constructed from a one-liter stainless steel beaker with a vacuum plate acting as its lid. All major suppliers sell these beakers and most sell vacuum plates and gaskets that fit perfectly. Many restaurant supply houses carry the same name brand beakers at much lower cost.

Fume Hood. For many laboratories, acquisition of a fume hood is a major expenditure. While building codes vary, OSHA and other agencies have established standards that should be followed. These agencies have Web sites and personnel willing to assist with design questions.

Electronic Scales. Electronic scales, capable of measuring in milligrams, are also an essential tool. Many of these instruments can be zeroed after each chemical addition, making the measurement process easy and more accurate. Be sure to account for the full range and volume of materials that will be used in the laboratory when choosing a scale.

Explosion-Proof Freezer. Not all the experiments in this book require the use of an explosion-proof freezer. The use of a freezer may be beneficial, however, when a treatment strategy calls for mixing even a small amount of catalyst into a Passivation Polymer solution. The catalyzation process is slowed appreciably when artifact processing is conducted in a freezer. A freezer unit is also helpful when a conservation strategy requires a slow, controlled acetone/Passivation Polymer exchange.

Small Necessities in the Laboratory

Disposable Cups. Disposable containers are practical when working with Passivation Polymer solutions. Before purchasing a large supply of any particular brand, however, test the container to be sure it does not react with solvents. While it is possible to clean glassware, paper cups and plastic containers that can be discarded after use are time savers.

Stir Rods. Plastic or glass stir rods work best with Passivation Polymers. *Never* use wooden stir sticks or tongue depressors to mix polymer/cross-linker solutions. The wood will absorb the chemicals differentially and interfere with the stoichiometry of the solution.

Tongs and Tweezers. Because artifacts come in all shapes and sizes, it is a good idea to have a variety of big and small tongs and tweezers in the lab. Restaurant supply companies are an excellent source for large stainless steel utensils. For specialty tweezers and implements for handling fragile materials, hobby shops and chemical supply companies generally have a better selection.

Acetone. Many processes outlined involve an acetone/Passivation Polymer exchange process. Acetone and other solvents are necessary for expedient, thorough exchange of water with polymer preservatives. Proper handling, storage, and waste management procedures are mandatory. Stainless steel or high-density polyethylene safety cans outfitted with spring-loaded safety caps are the safest and easiest way to handle solvents in the laboratory. These are available in a variety of sizes from chemical and safety supply houses.

Metal Screens and Vegetable Bags. For smaller artifacts, aluminum window screen can be friction-fitted on top of the artifact in solution to prevent them from floating during acetone/polymer exchange processes. Mesh bags, such as those used for bagging perishable foods in a grocery store, can help keep small objects together during treatment. Because of their open mesh design, these bags do not impede treatment.

Absorbent Materials. Working with Passivation Polymer–treated artifacts can be messy unless some simple precautions are taken. After treatment, artifacts are generally allowed to drain of excess free-flowing oils while sitting on screen material. Since these materials are often reusable, it is advisable to save runoff Passivation Polymers. After draining, however, the surfaces of the artifact need to be wiped of excess Passivation Polymer solution. Layers of lint-free paper and, in some cases, newspaper can facilitate cleanup.

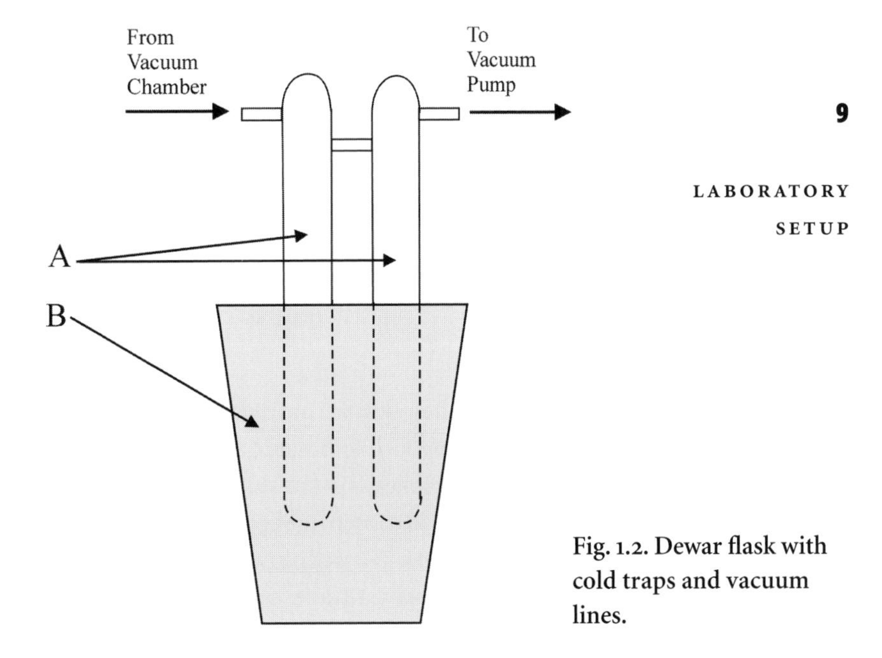

Fig. 1.2. Dewar flask with cold traps and vacuum lines.

Chemicals

Conservators working with Passivation Polymers to preserve organic artifacts need access to a variety of chemicals, but they must keep in mind important considerations in selecting and safely storing chemicals for use in a laboratory. Hundreds of silicone oils, cross-linkers, and catalysts are available for use in an archaeological chemistry laboratory. The most important consideration in selecting polymers is the safety of laboratory staff and students. Passivation Polymers were selected for developmental research at Texas A&M University because they are relatively safe to use and because research has proven

Table 1.1 Cross-Linkers

Catalyst	*Noted Attributes*
CR-20	Flexible stoichiometry Generally 3–5% addition by weight of the chosen polymer High % additions of CR-20 result in slightly brittle polymer structures
CR-22	Less flexible stoichiometry Generally 3% addition by weight of the chosen polymer Tends to create more brittle polymers

that they do not have detrimental effects when applied to organic artifacts.

Regardless of the relative safety of the chemicals chosen, the conservator should always work in a well-ventilated environment. Accordingly, a fume hood large enough to conduct day-to-day work is essential. While a fume hood may be the biggest investment when outfitting a laboratory, it is essential equipment. If used properly, the airflow created by a fume hood will ensure that contaminants and harmful vapors are not a problem.

Material safety data sheets must be available for all materials being used in the laboratory. These sheets contain the necessary information for lab personnel to work safely with chemicals. In the event of a spill, these sheets also contain guidelines for safe and controlled cleanup. Medical treatment information and emergency contact numbers should also be listed on the sheets. MSDS information is available on the Web. If the laboratory has a computer, it is advisable to have a "hot key" or posted instructions so that personnel can easily access safety information and emergency assistance.

Table 1.2 Catalysts

Material	*Application Method*	*Chemical Reactivity*	*Results*
CT-32 (DBTDA)	Wiped, painted	Long reaction time	Clear reaction
	Vapor deposition—containment chamber at elevated temperatures	Long reaction time (18 hrs)	Clear reaction
	Containment chamber at room temperature	Longer catalyst life	Clear reaction
CT-30	Wiped, painted	Shorter catalyst life than CT-32	Clear reaction
	Vapor deposition—containment chamber at room temperature	Shorter catalyst life than CT-32	Clear reaction
	Containment chamber at room temperature	Catalyst life not as good as CT-32	Clear reaction
CT-34	Wiped, painted	Very fast	Clear reaction
	Vapor deposition—containment chamber at elevated temperatures	Very fast—often slightly yellow coloration	Often a yellow coloration; often a powdery residue that can be brushed clean
	Containment chamber—not recommended	Oxidation occurs quickly	Often oxidation occurs without canalization taking place

CR-20 is the main cross-linker used for processing archaeological materials.[1] CR-20 is a hydrolyzable, multifunctional alkoxysilane capable of tying two or more polymer chains together. It has a specific gravity of 0.94 at 25°C and a viscosity of 1.00 centistoke (CST). Added to PR-10, KP80, PA Fluids, and 4-7041, CR-20 is an effective cross-linking agent.

CR-22 cross-linker is similar to CR-20. It is a tetra-functional alkoxysilane that has a specific gravity of 0.94 at 25°C and a viscosity of approximately 1.00 CTS. Observations suggest that artifacts treated with polymer solutions including CR-22 are slightly more malleable after treatment than those treated with CR-20. More testing is required, however, to determine stoichiometric considerations for use with organic artifacts. Table 1.1 provides an overview of CR-20 and CR-22.

CT-32, a dibutyltin diacetate compound, is the most commonly used catalyst for preserving organic artifacts.[2] This catalyst can be applied either as a topical application or through vapor deposition to cure the polymer-impregnated artifact. CT-32 is a tin-based catalyst that has a working life of approximately 24 hours. It is a good general purpose catalyst for use in the laboratory, with a specific gravity of 1.32 and a molecular weight of 351.01. Its chemical formulation is $C_{12}H_{24}O_4Sn$. Exposure to CT-32 fumes should be avoided as should direct contact

Table 1.3 Silicone Oils

Material	*Viscosity*	*Expected Result*	*Additive*	*Catalyst*	*Approximate Reaction Time*
PR-10	Thin	Rigid, easy penetration	CR-20	CT-32	24 hr
			CR-22	CT-30	15 hr
			CR-20 or CR-22	CT-34	20 min
PR-12	Thin	Rigid, easy penetration	CR-20	CT-32	24 hr
			CR-22	CT-30	15 hr
			CR-20 or CR-22	CT-34	20 min
PR-14	Medium	Less rigid, easy penetration	CR-20	CT-32	24 hr
			CR-22	CT-30	15 hr
			CR-20 or CR-22	CT-34	20 min
KP80	Very thin	Very rigid depending on density of organic substrate	CR-20	CT-32	24 hr
			CR-22	CT-30	15 hr
			CR-20 or CR-22	CT-34	20 min
PA Fluid	Very thin	Not as rigid as KP80 but firmer than PR-10	CR-20	CT-32	24 hr
			CR-22	CT-30	15 hr
			CR-20 or CR-22	CT-34	20 min
4-7041	Thin	Similar to PA fluids but may be more brittle	CR-20	CT-32	24 hr
			CR-22	CT-30	15 hr
			CR-20 or CR-22	CT-34	20 min

with skin. When used in artifact conservation, the catalyst remains chemically stable.

CT-30 is also a tin-based catalyst. CT-30 differs from CT-32 in that it has a working life of approximately 16 hours in a contained environment. CT-30 has a specific gravity of 1.25, a molecular weight of 404.88, and a chemical designation of $Sn(C_8H_{15}O_2)_2$.

The fastest acting of the three catalysts used for experimental research is CT-34. Also known as "Tyzor" TPT Titanate, it is a mixture of tetraisopropyl titanate (99%) and isopropyl alcohol (1%). CT-34 has a working time of approximately 3 hours in a contained environment. It is pale yellow in color, which is problematic when the catalyst is applied using vapor deposition. Through the process of catalyzation, a yellow powder can form on the surfaces of the artifact. In most cases, the thin layer of powder is easily removed with a soft bristle brush. Chemical reactivity of the catalyst can be accelerated by warming the solution. Reaction time can be slowed by keeping the artifact cooled in a freezer or refrigerator during catalyzation. Table 1.2 provides an overview of these catalysts.

The main silicone oils used for preserving organic materials in the Archaeological Preservation Research Laboratory are dimethyl siloxane, hydroxy-terminated polymers with up to 5% dimethyl cyclosiloxane added (see table 1.3). PR-10, the medium viscosity of the readily available silicone oils, is an odorless polymer with a specific gravity of 0.97 at 25°C and a viscosity of 13,500.00 CST. PR-12 is slightly more viscous than PR-10. It has a specific gravity of 0.97 at 25°C and a viscosity of 20,000.00 CST. KP80 (75.00 CST) is less viscous than PR-10 silicone oil and has a specific gravity of 0.97 at 25°C. PA Fluid is also a dimethyl siloxane, hydroxy-terminated silicone oil with octamethylcyclotetrasiloxane and decamethylcyclopentasiloxane added. PA Fluid has a viscosity of 5.0 CST and a specific gravity of 0.95 at 25°C. Another silicone oil, 4-7041, is a polydimethylsiloxane, hydroxy-terminated polymer, with a viscosity of 2–6 CST and a specific gravity of 0.97 at 25°C.

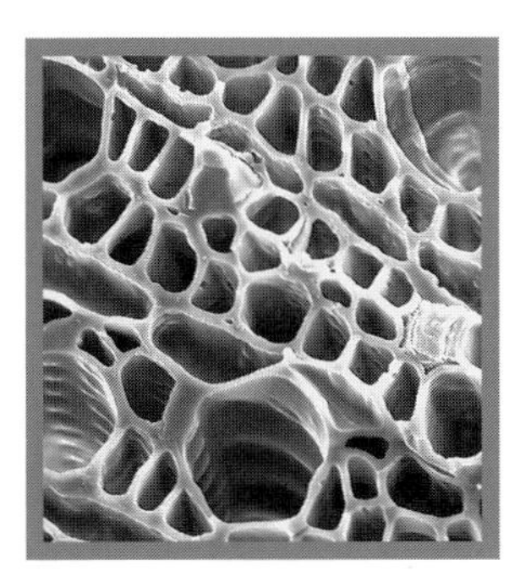

CHAPTER 2

Baseline Mechanisms

Silicone oils do not mix with water. This, then, is the first challenge a conservator faces after deciding to preserve waterlogged materials using polymers. Passivation Polymer artifact treatment involves three main steps: water/acetone displacement (WAD) or dehydration, acetone/polymer (AP) displacement or Passivation Polymerization, and catalyzation of the polymer-impregnated artifact.

Two experiments in this chapter demonstrate the relationship between dehydration of water from organic materials and the ability of silicone oils to penetrate into organic substrates. In "Dowel Experiment," the importance of WAE displacement is illustrated. Dehydration processes and length of time in dehydration influence penetration depth and rate of AP exchange in wooden dowels. The second experiment, "Mass Spectrographic Analysis of Out-Gases Created from the Dehydration of Archaeological Wood Samples," discusses single- and multiple-solvent dehydration techniques for organic materials and how multiple-solvent dehydration may enhance penetration depth.

Experimentation is ongoing to develop processes for determining nondestructive percentage dehydration of organic materials. Many processes are available for determining absolute dehydration in organic substrates. Unfortunately, most require testing methods that either damage the sample or are potentially stressful to the artifact in general.

Removing water from any waterlogged artifact can be harmful, especially if waterlogging has caused severe cellular damage. In many cases, water in the cells is all that is maintaining the diagnostic attributes of the artifact. The conservator must therefore understand the importance of proper dehydration as a precursor to treatment with any bulking agent, including Passivation Polymers. Pretreatment dehydration is generally not required for dry and desiccated organic materials. If wood or basketry feels damp due to absorption of atmospheric moisture, either dehydration of the samples or storage in a low humidity environment for a few days may be necessary to ensure the rapid uptake of polymers during treatment.

Dowel Experiment

The purpose of this experiment was to determine rates of acetone and silicone displacement into the matrix of groups of structurally solid wooden dowels that had undergone water/acetone displacement for one, two, three, or four days before treatment. Because of the tight, undamaged cellular structure of the dowels, this experiment represents a difficult conservation scenario for wood. The cellular structure of most waterlogged and desiccated timbers facilitates the AP process and shortens the time required to impregnate the wood. This experiment showed that (1) the penetration depth of acetone is dependent on time and vacuum processing, and (2) Passivation Polymers will only penetrate the dowels to the depth of acetone within the samples.

Forty-eight dowels, each four inches long and one inch wide, were prepared so that their cut ends could be sealed by dipping them in a tray of five-minute epoxy. Sealing the end surfaces ensured that all penetration of acetone and Passivation Polymers into the core of the dowels was through the sides of wood cells and not the tangential, porous ends of the dowels. After the epoxy had dried to form hard end caps, the dowels were placed into a five-gallon pail and immersed in fresh industrial-grade acetone. A section of aluminum screen and a five-pound weight were placed over the dowels to ensure that they remained submerged throughout treatment. The pail was then placed into a large vacuum chamber.

All the samples were dehydrated for 24 hours at room temperature and a reduced pressure of 5333.33 Pa (40 Torr). Twelve samples were removed from the pail daily and placed into a beaker of PR-10 silicone oil solution. While the first samples were being analyzed, the remainder of the dowels continued in dehydration so that day-4 dowels had four times as much dehydration and, presumably, deeper penetration of acetone into the core, than did day-1 dowels. (Day-1 dowels received 24 hours dehydration; day-2, 48 hours; day-3, 60 hours; day-4, 72 hours.)

Each day, the dowels were immersed in PR-10 and then treated to continuous acetone/Passivation Polymer displacement for 24 hours at a reduced pressure of 5333.33 Pa (40 Torr). During this 24-hour period, two samples were removed from the Passivation Polymer solution after 0.5, 1, 2, 4, 8, and 24 hours. Hence, the time of dehydration and the time of treatment were the only variables in the experiment. At the designated times, dowels were removed from the beaker and quickly surface-wiped with a paper towel. Each dowel was labeled and then promptly cut in half on a band saw. The wet line of Passivation Polymer penetration could be seen and marked with a ballpoint pen on the crosscut surfaces of each half of the dowel. It was important to mark this penetration depth (PD) as quickly as possible because, in many cases, this line of penetration diffused quickly after the samples were cut. Figure 2.1 shows a typical cross-sectional view of Passivation Polymer penetration of a dowel.

In this experiment, it was reasoned that as long as all samples were processed in the same manner and within the same time frame, the data should accurately represent the PD for each sample and the differences in PD between samples. The mean PD was then calculated for each collection time. After four days, four series of dowels had undergone the same processes of acetone/Passivation Polymer displacement and the same assessments for PD. Data for PD among all the dowels for all four days could then be compared (see tables 2.1–2.4). Dehydration time was expected to correlate with PD of acetone into the matrix of the samples. These same samples should also have a greater PD of Passivation Polymers into their matrix as compared to samples treated with a shorter period of acetone dehydration.

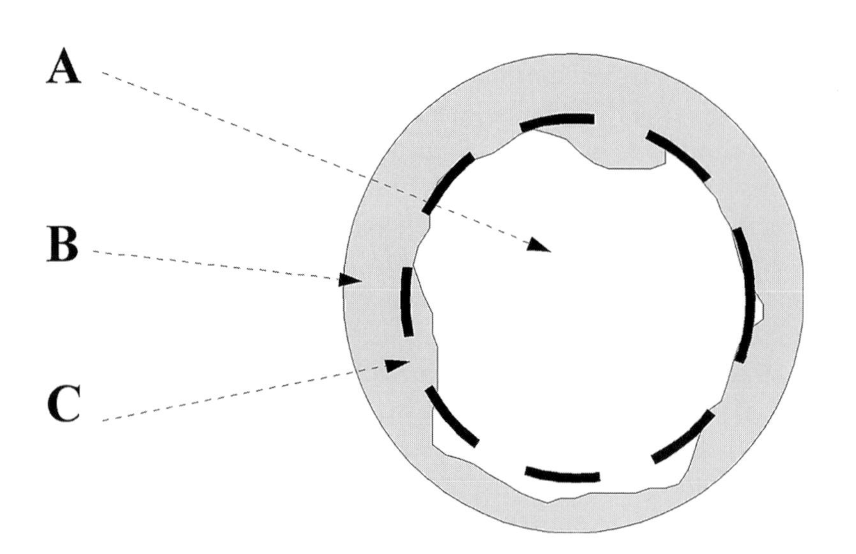

Fig. 2.1. Cross-sectional view of dowel and penetration depth measurement: *(A)* dry core; *(B)* Passivation Polymer wet zone; *(C)* measured mean penetration depth.

Table 2.1 One-Day Dehydration Data

Sample Number	*Treatment Time (in hrs)*	*Penetration Depth (in mm)*	*Mean Penetration Depth (in mm)*
8.1.1	0.5	1.75	2.875
8.1.2	0.5	4.00	
8.2.1	1.0	4.00	4.000
8.2.2	1.0	4.00	
8.3.1	2.0	5.00	5.000
8.3.2	2.0	5.00	
8.4.1	4.0	6.50	6.750
8.4.2	4.0	7.00	
8.5.1	8.0	6.50	6.500
8.5.2	8.0	6.50	
9.6.1	24.0	7.25	7.375
9.6.2	24.0	7.50	

Table 2.2 Two-Day Dehydration Data

Sample Number	*Treatment Time (in hrs)*	*Penetration Depth (in mm)*	*Mean Penetration Depth (in mm)*
9.1.1	0.5	5.50	5.500
9.1.2	0.5	5.50	
9.2.1	1.0	5.50	5.750
9.2.2	1.0	6.00	
9.3.1	2.0	6.50	6.750
9.3.2	2.0	7.00	
9.4.1	4.0	7.00	6.750
9.4.2	4.0	6.50	
9.5.1	8.0	7.00	7.250
9.5.2	8.0	7.50	
9.6.1	24.0	8.00	8.375
9.6.2	24.0	8.75	

Table 2.3 Three-Day Dehydration Data

Sample No.	*Treatment Time (in hrs)*	*Penetration Depth (in mm)*	*Mean Penetration Depth (in mm)*
10.1.1	0.5	5.50	5.750
10.1.2	0.5	6.00	
10.2.1	1.0	7.00	7.000
10.2.2	1.0	7.00	
10.3.1	2.0	7.50	7.500
10.3.2	2.0	7.50	
10.4.1	4.0	7.75	7.875
10.4.2	4.0	8.00	
10.5.1	8.0	8.00	8.500
10.5.2	8.0	9.00	
10.6.1	24.0	8.50	8.500
10.6.2	24.0	8.50	

Table 2.4 Four-Day Dehydration Data

Sample No.	*Treatment Time (in hrs)*	*Penetration Depth (in mm)*	*Mean Penetration Depth (in mm)*
11.1.1	0.5	5.00	5.000
11.1.2	0.5	5.75	
11.2.1	1.0	6.50	6.500
11.2.2	1.0	6.75	
11.3.1	2.0	7.00	7.250
11.3.2	2.0	7.50	
11.4.1	4.0	8.00	8.250
11.4.2	4.0	8.50	
11.5.1	8.0	8.50	8.500
11.5.2	8.0	8.50	
11.6.1	24.0	8.75	8.750
11.6.2	24.0	8.75	

Data for Dowel Experiment

The data were analyzed by multiple linear regression. As shown in figure 2.2 PD bore a logarithmic relationship to the hours the samples were in Passivation Polymer solution; the data were transformed by taking the logarithms of the number of treatment hours. The analysis showed PD depended upon the days of dehydration ($p < .001$) and the log-transformed hours of treatment ($p < .001$). One outlier was seen; its exclusion did not affect the results' significance. Hence, Passivation Polymer penetration into the matrix of these wooden dowels was directly linked to the penetration depth of acetone in the samples. Water/acetone exchange dehydration and acetone pretreatment of samples that appear to be dry play an important role in the bulking process of organic materials. Acetone and other water miscible solvents act as a conduit for the exchange of Passivation Polymer solutions into the core of an organic specimen.

In all cases, the degree of silicone depends upon the amount of water/acetone exchange that has occurred throughout the specimen being treated. This certainly dictates that, for most materials, the time spent in water/acetone exchanges cannot be bypassed. Indeed, a good, simple test is needed to indicate when acetone has completely penetrated the structure being treated and as much water as possible has been removed from the specimen.

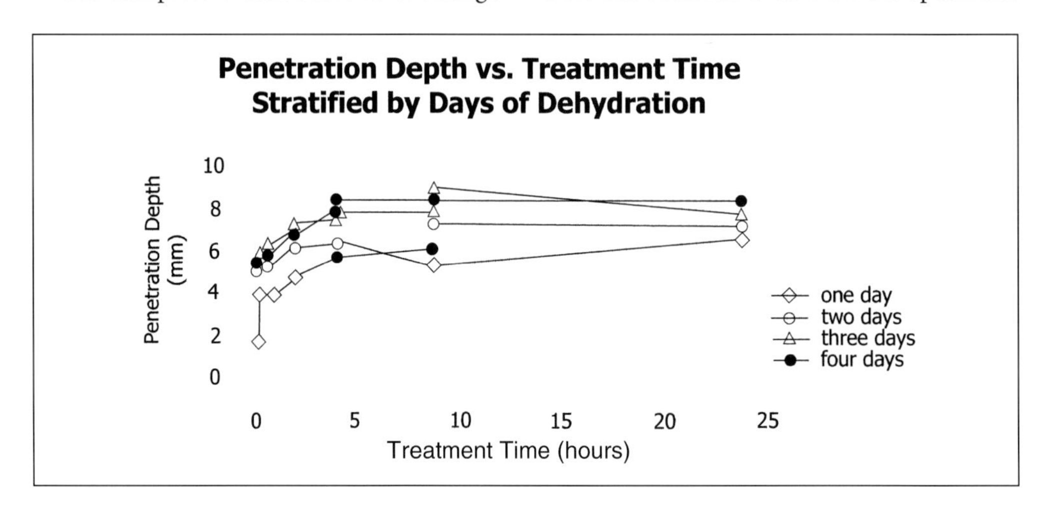

Fig. 2.2. Penetration depth vs. treatment time stratified by days of dehydration.

Mass Spectrographic Analysis of Out-Gases Created from the Dehydration of Archaeological Wood Samples

During development of Passivation Polymerization processes, the degree of dehydration and the methods used for dehydration of organic materials were discovered to affect treatment results. Artifacts treated using water/ethanol exchange (WEE) and ethanol/polymer exchange (EPE) processes produced very different results from similar materials preserved using WAD/AP processes. Mass spectrographic analysis was used to study the out-gases from similar size pieces of waterlogged wood that were dehydrated using WEE and WAD exchange processes.

Two factors are important for conserving organic materials with silicone oils. While it is not necessary to eliminate all water from an artifact, it is essential to remove as much of the free-moving water as possible. Acetone can easily be driven from samples in a vacuum chamber, so a dehydration process that effectively causes a higher output of volatile vapors should also facilitate a more effective exchange of silicone oils to replace the acetone. To evaluate which dehydration processes are best suited for conservation practices utilizing silicone-bulking techniques, a test was devised to study collected gases from dehydrated samples. Electron ionization mass spectrographic analysis was performed on the out-gases from waterlogged archaeological wood. The five most commonly used dehydration processes at the Preservation Research Laboratory are represented in table 2.5.

Electron Ionization Mass Spectral Analysis

Electron ionization (E.I.) mass spectra of the out-gases from the five waterlogged wood samples were acquired on a VG Analytical 70S high-resolution, double-focusing magnetic sector mass spectrometer. The VG 70S was equipped with a VG 11/250J data system that allowed computer control of the instrument, data recording, and data processing. The E.I. source was maintained at 220°C during the analyses, while 70 eV positive analyte ions were extracted and accelerated to 8 keV and then mass analyzed.

Gaseous samples were collected in a one-liter glass bulb and allowed to leak into the mass spectrometer through a regulated inlet. Liquid samples were analyzed by injecting 2 IL of the sample into a 180°C heated reservoir. The sample was then allowed to leak into the mass spectrometer through a glass jet. Approximately 15 mass spectra were taken and averaged per sample; then the mean ion abundance per sample was reported. Background measurements of water (m/z+18) were determined to be approximately 1 mV at the amplification used to collect the analyte mass spectra.

The first method of dehydration was to simply immerse the waterlogged wood in a series of fresh acetone baths, allowing adequate time for each bath to draw free-flowing water from the sample. The acetone used

Table 2.5 Commonly Used Dehydration Processes

Dehydration Process	*General Procedure*
Acetone	Series of baths of fresh acetone to displace water
Alcohol/acetone	Series of alcohol baths, followed by a series of baths in acetone
Acetone/vacuum in freezer	Series of acetone baths under vacuum at 0°C
Alcohol/acetone/vacuum in freezer	Alcohol and acetone baths at 0°C
Acetone/warm out-gassing	Series of acetone baths, followed by warming the artifact to 62°C to drive off water and acetone

was an industrial-grade solvent, which had been stored in a 55-gallon drum for approximately three months prior to use in this experiment. Because of the small sample size, sample A was placed into a beaker containing 500 ml of fresh acetone and allowed to remain at room temperature for 24 hours. The sample was then placed into an identical volume of fresh acetone for an additional 24 hours.

After this period of dehydration, the sample was removed from the acetone bath and lightly surface-tamped with a dry paper towel to remove excess acetone from the surface. Additional acetone was allowed to out-gas by placing the sample on a dry paper towel for two minutes. The sample was then placed into a clean beaker containing 300 ml of PR-10 silicone oil. The beaker was immediately connected to an in-line gas trap assembly with an attached vacuum pump (fig. 2.3). Dry ice surrounding the trap assembly was used to freeze the entire range of gases being given off by the sample under a vacuum of 28 Torr. After 20 minutes of applied vacuum, the contents within the gas trap were collected and placed into sterile specimen vials for transport to the mass spectrometer.

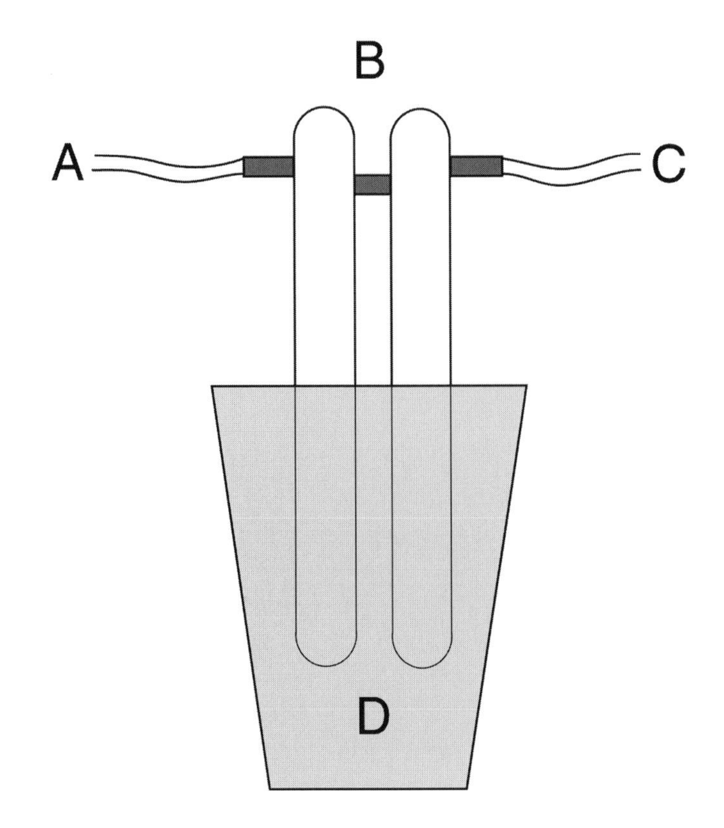

Fig. 2.3. Beaker and gas trap assembly mounted in dry ice: *(A)* vacuum line from vacuum chamber holding artifact; *(B)* two gas traps in series between the vacuum chamber and the vacuum pump; *(C)* vacuum line leading to vacuum pump; *(D)* Dewar flask, holding the gas traps and dry ice.

The second waterlogged wood sample, sample B, was dehydrated in much the same fashion as sample A, except that, for initial dehydration, it was placed into a beaker containing 500 ml of ethanol and allowed to stand for 24 hours at room temperature. The alcohol was then replaced with 500 ml of fresh acetone. After being capped to prevent evaporation, the sample was allowed to sit at room temperature for an additional 24 hours. Subsequently, sample B was removed from the bath and treated identically to sample A.

Waterlogged wood sample C was dehydrated using a series of acetone baths that were placed in a freezer-mounted vacuum chamber as a means of determining if the combination of low temperature and vacuum had a significant effect on the dehydration process. Sample C was placed into 500 ml of fresh acetone and placed into the freezer-mounted vacuum chamber, where it remained under a vacuum of 28 Torr for 24 hours. After 24 hours, the wood sample was quickly transferred to a clean beaker containing 500 ml of fresh acetone and then returned to the freezer vacuum chamber for an additional 24 hours of dehydration using the same vacuum pressure as before. Upon completion of the second acetone bath, the sample was removed from the freezer and surface-patted using dry paper towels. Subsequently, sample C was treated identically to samples A and B.

A fourth sample, D, received the same treatment as sample C except that, instead of two acetone baths under vacuum, initial dehydration was started using ethanol so that, after 48 hours, the sample had been dehydrated in one bath of alcohol and one bath of acetone. After collecting out-gases from the sample using the same apparatus and methodology used for the other samples, the vial was ready for analysis. (If a sample could not be immediately analyzed, it was stored in a freezer.)

Because silicone oil is nonvolatile and chemically pure, it seemed to be an ideal inert substance in which to immerse the wood samples during the out-gassing process. To test this hypothesis, an additional wood sample, E, was dehydrated in two successive baths of 500 ml of fresh acetone at room temperature. The sample was then placed into a clean beaker, where it was warmed by a heat lamp at 62°C until all the surfaces of the wood were light brown in color (fig. 2.4). At this point, simply touching the outer surfaces of the sample indicated that it was dry, although the weight indicated that the interior was still heavy with liquid. As with the other samples, out-gases from sample E were collected in a dry ice–cooled gas trap for a period of 20 minutes. After the products of out-gassing from sample E were placed into a sterile, sealed vial, it, too, was analyzed.

Results

No attempt was made to evaluate the physical appearance of the five samples used in this experiment. Variability among samples is another uncontrollable aspect of wood research. Although samples were all cut to the same size for testing (3.0 cm long, 2.0 cm wide, and 0.5 cm thick), differences in wood growth, degree of waterlogging, fungal and bacterial degradation, and a host of other factors can expand the degree of variability among samples. Data from mass spectrographic analysis for each sample, however, answered the basic questions of water replacement and mobility of volatile solvents within the samples. The data showed significant differences among the relative intensities of elements collected during experimentation.

Sample A was dehydrated using a sequence of two room temperature acetone baths, each lasting for 24 hours. Relative intensities of water, nitrogen, methyl, and oxygen in this sample were the highest for all samples tested, with amounts recorded as 2.5, 6.5, 25, and 2.0, respectively. The intensity of molecular acetone in this sample was approximately equal to samples B, C, and D. The levels of molecular acetone were nearly double in sample E, which was warmed during out-gassing.

Sample B was initially dehydrated at room temperature using a 24-hour alcohol bath, followed by an acetone bath. Recorded levels of nitrogen were approximately 50% less than those recorded for sample A, while the levels of methyl and ethyl were reduced by approximately 66% and 50%, respectively. Acetone levels in sample B were only slightly lower than those recorded for sample A.

Cold dehydration using successive baths of acetone and an applied vacuum to dehydrate sample C produced some interesting results. Analysis of the out-gases indicated a nearly twofold increase in methyl levels, while nitrogen levels for this sample were the lowest of all samples tested, recording an intensity of 2.5.

Sample D, which was dehydrated in a sequence of cold alcohol and acetone baths under vacuum, registered increased levels of methyl and nitrogen (19.6 and 3.8 intensities, respectively). Notably, the levels of acetone out-gassing were substantially greater (32.0 intensity).

Sample E was warmed during the out-gassing process. Silicone oil was not used to support the sample during the gas collection process. Levels of volatile elements such as methyl, ethyl, and acetone were predictably higher in sample E, although water out-

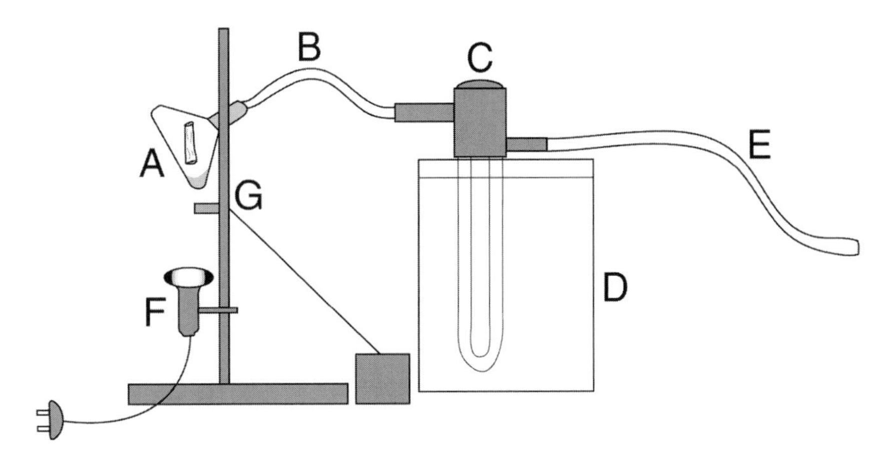

Fig. 2.4. Basic beaker and gas trap setup with heat lamp and temperature probe added: *(A)* beaker with wood sample—no oil; *(B)* vacuum line to gas trap; *(C)* gas trap mounted in dry ice; *(D)* container with dry ice; *(E)* vacuum line to vacuum pump; *(F)* heat lamp—adjustable height; *(G)* temperature probe.

Table 2.6 Data from Mass Spectrographic Analysis of Waterlogged Wood

Sample	*Nitrogen*	*Methyl*	H_2O	C_2H_3	*Ethyl*
A	6.50	25.00	2.50	7.50	2.50
B	3.00	8.00	2.00	13.00	14.00
C	2.50	14.80	2.00	12.00	2.50
D	3.80	19.60	2.00	6.00	2.50
E	4.00	22.50	2.00	8.00	3.80
Sample	*Oxygen*	C_3H_3	*Acetone*	*Acetone*	*Acetone*
A	2.00	4.50	100.00	2.00	28.00
B	2.00	4.50	100.00	3.00	25.00
C	4.50	7.50	100.00	2.40	24.30
D	0.20	5.00	9.00	100.00	32.00
E	1.80	3.50	12.50	100.00	53.00

gassing from the sample was comparable to samples B, C, and D. A lack of media to support the degraded cell structure resulted in greater damage to this section of waterlogged wood. Not surprisingly, the levels of out-gassing methyl and ethyl were the highest in this sample, and molecular acetone levels were 65% (53.0 intensity) higher than those recorded for sample D.

Trends

Sample E, which acted as a control for this experiment, served to indicate the maximum out-gassing potential that might be expected for this group of wood samples. Because heat was used to accentuate the effects of out-gassing, methyl, ethyl, and acetone levels for the sample were the highest recorded. While less than the levels recorded for sample E, the levels for sample D were appreciably higher than out-gassing levels recorded for samples processed at room temperature.

A comparison of the relative levels of gases from samples C and D is interesting because the higher levels of methyl and acetone gases suggest that alcohol-acetone process dehydration may have been more effective in raising the overall levels of volatile vapor exchange. Thus this process may have a higher potential for exchanging silicone oils, or other bulking agents, into waterlogged materials.

With the exception of sample A, which registered water levels of 2.5, samples registered water levels of 2.0, which may suggest that, from the standpoint of the removal of free water from waterlogged wood, all processes were relatively equal. Differences in bulking potentials, then, lie in the heightened levels of volatile gases that result in cold process, vacuum assisted alcohol-acetone dehydration. Table 2.6 includes the out-gassing data for all five samples. The data suggest that the rise in molecular acetone and methyl levels resulting from multiple-solvent dehydration makes two-stage cold dehydration potentially more appropriate for silicone bulking.

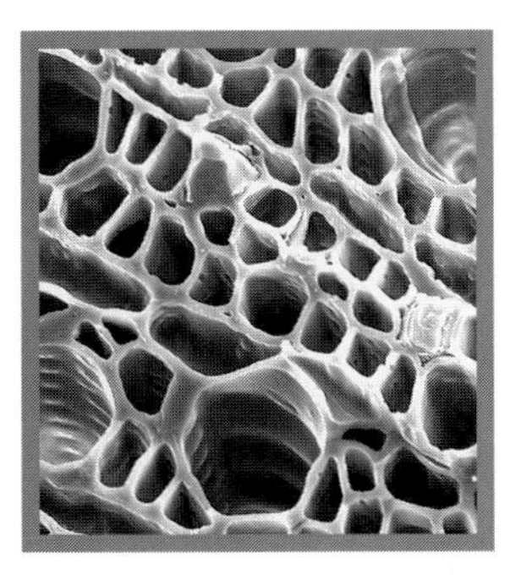

CHAPTER 3

Archaeological Wood

One of the biggest problems conserva tors may encounter when developing a conservation strategy for waterlogged wood is determining the degree of deterioration that has occurred within the matrix of the artifact. Whether from a bog, a river, or a swamp, a piece of wood will suffer structural deterioration as a result of bacterial decay and fungal and microbial attack. Effects of water flow and natural abrasion from sands and silt also compromise the physical integrity of wood if deposition has occurred in shallow water.

The Challenge of Conserving Waterlogged Wood

While it is safe to assume that it is easier to penetrate the matrix of heavily waterlogged wood with a bulking agent, to use the same bulking material for all artifacts would be too simplistic. Indeed, in the case of the large oak timbers in the hull of *La Belle,* the penetration rate of a bulking agent into the outer, heavily waterlogged timbers, will be fast. The inner core of wood, however, is extremely solid, exhibiting nearly the same strength and density as freshly sawn timbers. Penetration of a bulking agent into these timbers is much more difficult. As formidable as the task may be, however, the introduction of a stabilizing and consolidating agent is necessary to prevent exfoliation that occurs with the separation of the outer zone of wood from the denser interior. Passivation Polymers seem to effectively prevent this problem.

The use of silicone preservation technologies also allows the conservator to better control the strength, flexibility, exterior finish, and color when treating waterlogged or severely desiccated wood. As with polyethylene glycol, Passivation Polymers can be blended to accentuate structural and aesthetic qualities that the conservator may wish to apply to archaeological wood.

Because silicone oils do not mix with water, all free-flowing water must be removed to allow impregnation of the wood. This can be easily accomplished with solvent baths. To prevent stress, a series of baths should be used, starting with alcohol dehydration and working through to two or more acetone baths. For small artifacts, reduced pressure applied to the wood in solution will facilitate water removal. Care must be taken in applying a vacuum to acetone as vapors will contaminate the vacuum pump oil, and this can hinder performance of the equipment. Acetone will also damage a pump with plastic components. Older vacuum pumps are invaluable in a conservation laboratory because their all-steel construction is impervious to contamination. A Dewar flask with dry ice should always be used to house two inline gas traps between the pump and the vacuum chamber. This is the safest way to prevent acetone vapors from entering the pump.

Degradation and Shrinkage

After long periods in wet soil, peat bogs, or marine sites, the cell wall components in all wood will be degraded by bacterial action. In general, water-soluble substances such as starch and sugar are the first to be leached from waterlogged wood, along with mineral salts, coloring agents, tanning matters, and other bonding materials. In time, through hydrolysis, cellulose in the cell walls disintegrates, leaving only a lignin network to support the wood. Even the lignin will break down over a long period of time.

As a result of the disintegration of cellulose and lignin, spaces between the cells and molecules increase, and the wood becomes more porous and permeable to water. All the deteriorated elements of the wood, including cell cavities and intermolecular spaces, are filled with water. The remaining lignin structure of wood cells and the absorbed water act to preserve the shape of the wood. The loss of the finer cellulose tissue does not cause much alteration in the gross volume of wood, but the porosity is increased and the wood absorbs water like a sponge.

A waterlogged wooden object will retain its shape as long as it is kept wet. If the wood is exposed to air, the excess water evaporates and the resulting surface tension forces of the evaporating water cause the weakened cell walls to collapse, creating considerable shrinkage and distortion. The amount of shrinkage is dependent upon the degree of disintegration and the amount of water present. The amount of water in waterlogged wood can be determined by the following equation:

$$\%H_2O = \frac{\text{weight of wet wood} - \text{weight of oven dried wood}}{\text{weight of oven- dried wood}} \times 100$$

Wood containing more than 200% water is considered degraded; it is not uncommon to find wood that contains more than 500% or even 1000% water. Waterlogged wood is often classed according to the amount of water it contains. The most commonly used classifications are shown in table 3.1.

Table 3.1 Classifications of Waterlogged Wood

Classification	*Water*
1	More than 400%
2	185–400%
3	Less than 185%

All waterlogged timbers must remain completely wet up to, and throughout, the preservation process. Wood recovered from saltwater environments requires extensive rinsing in freshwater to eliminate as many soluble salts as possible. Failure to eliminate salts in the matrix of the wood will cause problems once the artifact is removed from treatment. Trapped salts will slowly crystallize, resulting in possible distortion and exfoliation of large layers of the artifact. Titration tests should be used to determine the point at which the bulk of soluble salts have been rinsed from the wood. As the results from the "Dowel Experiment" in chapter 2 suggest, Passivation Polymers penetrate easily into an acetone-rich environment. Accordingly, prior to treatment, cleaned wooden artifacts should be stored in an alcohol/acetone (50:50) environment. In many cases, artifacts can be stored in 100% acetone—the longer the better.

Waterlogged Wood from Saltwater Environments

This procedure deals with any wood with an internal structure that has been degraded by the waterlogging process.

Pretreatment

All waterlogged timbers must remain wet prior to treatment. Regardless of whether they were recovered from saline or freshwater environments, the water in the cell structures of these artifacts is preventing collapse and distortion. Any amount of air-drying or dehydration will cause potentially irreversible cellular collapse.

There are additional challenges in the treatment of waterlogged timbers from saltwater environments. With evaporation, salts remaining in the wood will crystallize. Expansion and crystallization as the artifact dries may result in extensive cracking and, possibly, exfoliation of large sections of wood. Even trace amounts of salts can damage structurally deteriorated wood. Because they remain soluble, salts within the cell structure of the artifact act to absorb atmospheric moisture and are not completely stabilized. Soluble salts and minerals must, therefore, be flushed from waterlogged wood prior to treatment. The best means of removing soluble salts is through freshwater rinses and, when possible, running water rinses. Over time, salts will dissolve into and be diluted in the volume of water in which the wood is stored. Accordingly, large volumes of water are best for long-term storage. Water should be changed frequently. Titration readings and simple silver nitrate tests can be used to determine parts per million (ppm) salt content prior to changing rinse water.

Vats containing waterlogged wood should be kept covered. Exposure to sunlight will encourage the growth of algae, which may be detrimental to the wood. Covering the vats will also help eliminate deposition of airborne contaminants, as well as mosquito and other insect larvae. How biocides affect the bulking agents used to stabilize waterlogged wood or the cell wall structures of the wood itself is not known. Accordingly, biocides should be used sparingly.

Methodology

1. The surfaces of the wood should be thoroughly cleaned using soft brushes, lint-free cloths, and other appropriate tools. Calcareous deposits can often be removed using dental picks or soft wooden dowels. Small pneumatic chisels are useful for removing slightly larger calcareous deposits. Care should be taken, however, to ensure that vibrations from the chisel do not shake the wood to the point of causing additional stresses. To minimize damage caused by vibrations, the artifact should be held in one hand or placed on soft vermiculite-filled cloth bags while removing concreted materials.

2. To enable deep penetration of polymers throughout the wood, water must first be eliminated. For small artifacts, this can be accomplished using a series of dehydration baths, starting with ethanol and working through to several baths of fresh acetone. Water/acetone exchange will take several days, depending upon the artifact's size and density. *This phase of treatment cannot be hurried.* Incomplete dehydration or dehydration carried out too quickly may damage the artifact.

Vacuum-assisted dehydration for the final acetone baths is often an effective way to ensure complete dehydration. During dehydration, the vat containing the artifact in fresh acetone is placed into a vacuum chamber. Two gas traps, immersed in dry ice or liquid nitrogen, should be placed in-line between the vacuum chamber and the vacuum pump. Escaping acetone fumes and moisture in the wood will be condensed and trapped in the gas traps, preventing contamination and possible deterioration of the internal components of the pump. Generally, reduced pressure using a slight vacuum of 5333.33 Pa (40 Torr) is sufficient to ensure thorough dehydration. Once dehydration is complete, the artifact must be transferred to the polymer solution while thoroughly wet with acetone.

3. A suitable volume of Passivation Polymers (silicone oil mixed with a cross-linker) should be prepared in advance to ensure quick transfer after dehydration. A

Passivation Polymer can be selected from table 1.3. Because waterlogged wood is fragile, care must be taken to ensure that the chosen polymer/cross-linker solution is not so viscous as to prevent rapid acetone/polymer exchange. The use of viscous polymer solutions will have the same effect as using too high an initial percentage mixture of polyethylene glycol in water. If the chosen bulking agent cannot readily displace acetone in the cell structure of the wood, cellular collapse and distortion of the artifact may occur.

For wood that has a soft exterior zone surrounding a harder and possibly structurally sound core, it may be necessary to use a blend of Passivation Polymers. Lower viscosity Passivation Polymers will penetrate the tight, solid matrix of less structurally damaged core wood, while more viscous Passivation Polymers will easily penetrate the softer exterior. Because higher viscosity polymers, such as PR-12 and PR-14, are longer polymer chains, they are excellent consolidators, binding the outer soft areas of waterlogged wood to the more solid core. If the PEG is too viscous, cell walls will be collapsed as the solution tries to enter the cells. Similarly, an overly viscous solution of silicone oil will cause cell collapse. Acetone/silicone oils solution displacement will have the same damaging effects as PEG if the solution tries to displace acetone, which is in this case dispersed rapidly as water.

Once a suitable volume of polymers or blended polymers, sufficient for total immersion of the artifact, has been selected, a cross-linker must be added. CR-20 is the best choice of cross-linking agents to use with PR-10, PR-12, and PR-14 for preserving waterlogged wood. A 3% (by weight) addition of CR-20 should be mixed in with the polymers.

4. After thoroughly mixing the polymer/cross-linker solution, quickly and carefully transfer the acetone-laden wood to the polymer solution. The artifact must remain immersed in the polymer solution during treatment. Throughout acetone/polymer solution displacement, the artifact may have a tendency to float as acetone vapors escape from the wood. It may be necessary to place aluminum screen and sufficient weight on top of the screen to keep the wood submerged (fig. 3.1).

5. To ensure the role displacement of acetone with a silicone oil solution, artifacts the size of a sewing thimble may require treatment for one week at ambient pressure. A certain degree of experience is needed to determine the best treatment strategy for small artifacts. If the artifact is in but not severly damaged by the waterlogging process, vacuum pressure may be used to assist the acetone/silicone oil solution displacement process. If however, the artifact is more severely damaged, it may be necessary to conduct the acetone/silicone oil solution displacement process at ambient pressure. At ambient pressure, acetone will vaporize more slowly from the cells of the wood, allowing the more viscous polymer solution slowly into the cells.

For larger artifacts such as musket stocks and tool handles the conservator may choose to use vacuum-assisted acetone/polymer displacement. Low viscosity polymers must be used for low-pressure treatment of wood. Accordingly, PR-10 is generally the best choice. To a given volume of PR-10 polymer, use a 3% (by weight) addition of CR-20 and mix thoroughly. The artifact should then be carefully transferred and totally immersed in the vat of polymer solution. Aluminum mesh or some other mechanism is required to prevent the artifact from floating during acetone/polymer displacement. Once the artifact is secure in the solution, place the vat into a vacuum chamber. A slight reduction in pressure is sufficient to cause small bubbles to flow from the artifact. When slight bubbling occurs, lock off the vacuum chamber and turn off the pump. Never allow extensive bubbling to occur as this means acetone is probably being driven from the wood faster than the polymer solution can displace it. Vacuum-assisted treatments work best when pressure is reduced slowly over several days. Generally, a reduced pressure of 5333.33 Pa (40 Torr) is sufficient for the treatment of any small wooden artifacts if the conservator has elected to use a reduced pressure environment to speed up the acetone/polymer displacement process.

Experience and good observations dictate the amount of vacuum pressure that can be applied to an artifact. If it is obvious that the diagnostic attributes of the artifact are being maintained, it can be held at pressure in solution from 12 to 18 hours. With badly degraded wood, acetone/polymer solution exchange is usually fast since the wood's deteriorated structure allows rapid and safe exchange of acetone for the more viscous polymer solution. For less degraded wood, exchange of acetone for polymers must be conducted at a slower rate to prevent distortion and cellular collapse.

6. After bubbling has ceased, the artifact should remain in a reduced pressure environment for several hours, if not days. Care should be taken to ensure that the reduced pressure in the vacuum chamber is returned to ambient pressure slowly. After the artifact has been allowed to sit in solution at ambient pressure for at least 24 hours, it can be removed from the Passivation Polymer solution. The artifact should then be carefully placed on a screen suspended over a clean container to allow the polymer solution flowing from its surfaces to collect. Recovered solution can be reused almost indefinitely, as long as it is periodically placed for 1 hour in a warming oven set at 30°C. This will ensure that remaining acetone and water are evaporated from the polymer solution. With each reuse, a new 3% addition of CR-20 is required.

The conservator need not be concerned about the safety of the wood while it is draining. It is best, in fact, to let the wood drain until its surfaces are nearly dry to the touch. This may take one or two days. To remove surface-pooled polymers from the wood, use lint-free cloths and soft brushes. A Q-tip dipped in CR-20 is often an ideal tool. To remove excess amounts of bulking agent from cracks and crevices, it may be necessary to immerse the artifact in a small vat of CR-20 and then wipe it with soft cloths and brushes.

7. A list of catalysts is supplied in table 1.2, which also lists the attributes of each catalyst and its intended end result. The best general-purpose catalyst is CT-32, which has a working time of 24 hours in a closed environment. Catalysts can be applied either topically or as vapors to the surfaces of artifacts. In either case, an applied catalyst will continue working to polymerize the polymer-impregnated artifact.

Topical applications are effective ways to initiate polymerization and attain total stabilization of the artifact. Use a soft brush or cloth to apply a thin, even layer of catalyst to the artifact's surfaces. After allowing the catalyst to coat the surfaces of the wood for two or three minutes, use soft cloths to wipe off the solution. Place the artifact into a Ziploc bag and seal it shut. After 24 hours, remove the artifact and inspect it to ensure that it is dry to the touch. If it is not dry, it may be necessary to return it to the Ziploc bag, along with a small piece of cloth containing several drops of fresh catalyst. Seal the bag again and allow fresh catalyst fumes to complete the catalyzation process.

A more aggressive form of vapor catalyst deposition can be accomplished by placing the artifact into a Ziploc bag or plastic container, along with a dish containing several grams of CT-32 catalyst (fig. 3.2). Seal the container and place it in a vented warming oven set at 34°C. Warming the catalyst will accelerate its reactivity with the polymer solution in the artifact. Generally, the artifact will be completely stable after 24 hours. When in doubt, you can store wood in

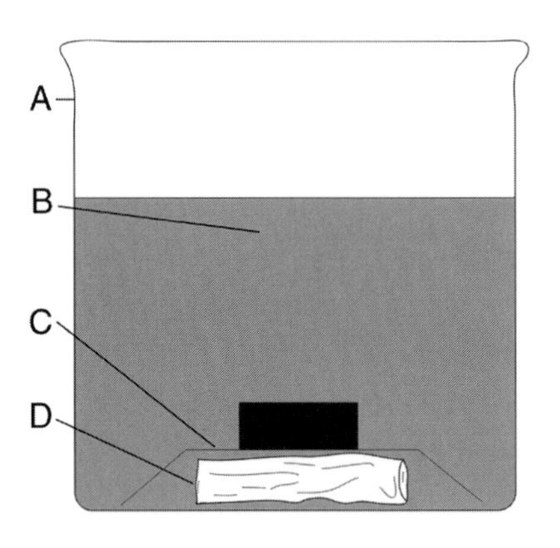

Fig. 3.1. Artifact in bulking agent with friction-fit screen in place: *(A)* open container; *(B)* sufficient bulking agent to submerge artifact; *(C)* friction-fit screen to ensure immersion during treatment; *(D)* artifact.

a Ziploc bag for several days, as long as fresh catalyst is added daily. After catalyzation, allow the artifact to sit in a fume hood for one or two days. The smell of the catalyst will dissipate quickly.

Case Study: Waterlogged Wooden Buttons with and without Associated Thread

After their recovery from the shipwreck *La Belle,* wooden buttons with and without associated thread were stored in freshwater. Initially tap water, with a chloride level of approximately 82 ppm, was used to desalinate the buttons. Over the course of one month, the bath water was changed every two days. For the next two weeks, the buttons were rinsed in baths of rainwater, which like the tap water baths were changed every second day. To complete the desalination process, the buttons were rinsed in five baths of deionized (DI) water, which were also changed every second day. Titration testing of the rinse water was conducted periodically throughout the desalination process. Generally, a level of 30 ppm or lower is considered adequate for starting the next phase of treatment.

Fig. 3.2. The artifact in a Ziploc bag with a dish containing several grams of CT-32 catalyst.

Dehydration of the buttons was conducted using a series of organic solvent baths. Starting with a solution of 25% ethanol in DI water, each successive bath included a 25% increase in ethanol until the buttons were dehydrated in 100% ethanol. Each phase of alcohol dehydration lasted 14 days. The solvent baths continued, with each successive bath receiving a 25% increment of acetone. Each bath lasted 14 days. The final stages of dehydration included three successive baths of 100% acetone, each lasting 14 days.

A solution of PR-10 with 5% CR-20 added by weight, sufficient to immerse the buttons, was prepared. To ensure that the buttons and attached threads did not float during treatment, small sections of aluminum window screen were friction-fit over them in the container (fig. 3.3). During acetone/polymer displacement, the solvent vaporizes and escapes from the wood, creating voids; continuous immersion of the buttons is therefore needed to ensure that these voids are rapidly filled with polymer solution. The buttons remained immersed in the polymer solution for 10 days.

After 10 days, the buttons were removed from the solution and placed onto a screen so that runoff polymers could be collected for reuse (fig. 3.4). Each button was then individually wiped, first with paper towels, to remove the bulk of polymers, then with soft cloths to remove any remaining solution. As each button was dried, it was immediately dropped into a bath of 100% CR-20 for a period of no longer than two minutes so that additional polymers could be washed from its surfaces. Each button was then removed from the CR-20 and carefully surface-dried, taking care not to dislodge any associated thread.

The buttons were then placed on an elevated window-screening tray in a Ziploc bag. A paper towel was crumpled into a ball and placed into the bag, along with the artifacts. The towel served to form an airspace around the artifacts. A second piece of crumpled paper, containing 250 cc of CT-32, was then placed into the bag (fig. 3.5). Once sealed, the bag was placed into a warming oven preset at 37°C. After 24 hours of catalyzation, the paper containing the used catalyst was replaced with a new ball of paper containing 250 cc of fresh CT-32. The bag was resealed and returned to the warming oven for an additional 24 hours.

Care must be taken to ensure that paper containing the CT-32 catalyst does not come into direct contact with any artifacts during catalyzation. Direct contact with the catalyst will result in the formation of white particulate at the point of contact. This is polymerized silicone oil, which can easily be removed using a soft cloth once treatment is complete. Avoiding contact, however, will eliminate the need for additional cleaning. After 48 hours of vapor deposition catalyzation, the buttons felt both dry and nonwaxy. They were then removed from the Ziploc bag and placed on a tray for continued curing in a fume hood.

The final appraisal of the treatment process for each button was subjective. If the buttons had felt waxy or had appeared to be wet, they could have been immersed in CR-20 to surface-clean the slight polymer film that may have formed during treatment. All the buttons with attached threads were dipped in CR-20 and then carefully surface-dried. This made the threads more flexible.

Fig. 3.4. Buttons draining on a screen after polymer/cross-linker treatment.

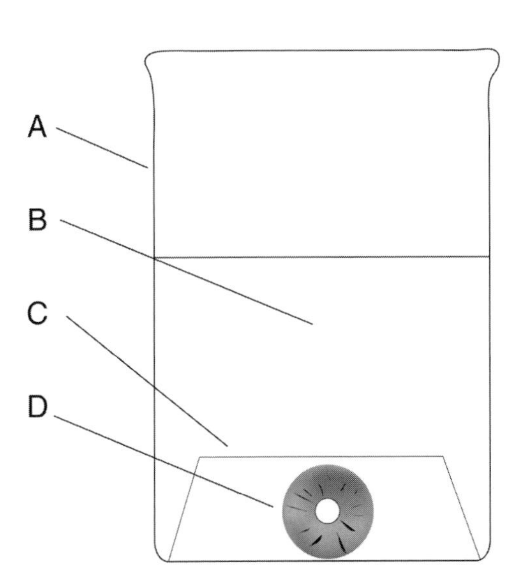

Fig. 3.3. Button immersed in polymer/cross-linker solution: *(A)* beaker; *(B)* polymer/cross-linker solution; *(C)* aluminum screen; *(D)* button.

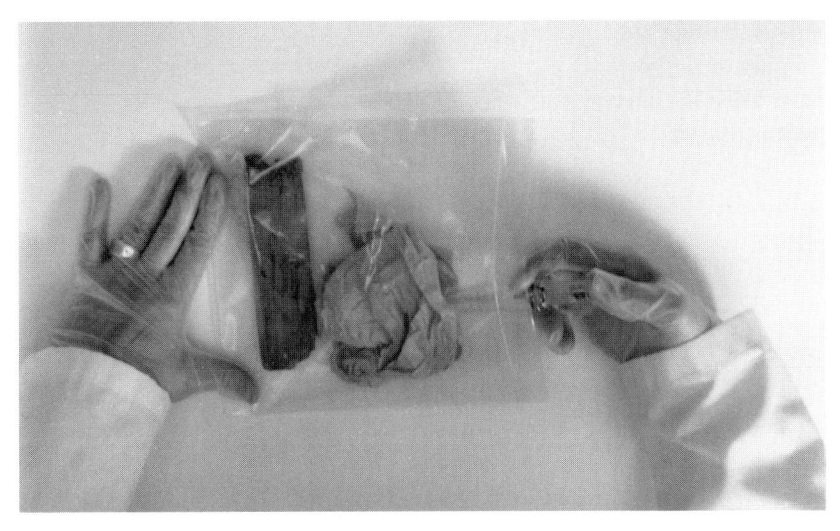

Fig. 3.5. Buttons being prepared for the vapor catalyst deposition process.

Inspection revealed the individual strands of the threads were well preserved and not clumped together. Any threads that appeared to be slightly clumped were easily separated while pat-drying the CR-20. Over time, the threads have not become stiff and have delighted conservators visiting the lab (fig. 3.6).

The results of Passivation Polymer treatment of wooden buttons, with and without associated threads, have been very successful. The artifacts retain their diagnostic features with only 1.5–2.0% shrinkage of overall size. The associated threads have retained their color, yarn memory, flexibility, texture, and some elasticity. Over time, the artifacts have shown no change in their original posttreatment appearance or condition.

Dry-Site Artifacts—Dry and Desiccated Wood

This procedure deals with weathered wood and architectural materials, and with organic artifacts from dry cave sites.

Pretreatment

Usually, dry and desiccated wood samples recovered from land sites are fragile and light in weight due to water loss. Over time, the lignum, starches, polysaccharides, polyphenolics, and water within the matrix of an artifact are lost through desiccation, leaching, oxidation, or attack by fungus and dry rot. Excessive handling in this compromised state is difficult, and achieving long-term stability may be impossible.

The use of polymers for the treatment of dried wood is a simple process since dehydration is seldom required. Completely eliminating water from wood prior to treatment using Passivation Polymers is not necessary. Artifacts that have been exposed to a dry environment or that have been stored in an environment of less than 50% humidity seldom need posttreatment dehydration. For small artifacts, such as buttons, the following processes are effective conservation strategies.

Methodology

Where possible, the surfaces of all artifacts should be cleaned of loose debris. Most of the time, this can be accomplished with a soft brush or lint-free cloths. In situations where dirt is heavily impacted on an artifact, however, it may be necessary to dampen the surfaces before attempting any cleaning. After cleaning, sufficient time should be allowed for the artifact to dry before continuing treatment.

While the artifact is air-drying, prepare a suitable volume of silicone oil/cross-linker solution to ensure total immersion of the assemblage throughout treatment. Be sure to consider the potential volume of polymer solution that will be absorbed during treatment. Because dry and desiccated wood tends to be friable, choosing a suitable polymer/cross-linker combination is important. Absorption of the polymer will be rapid in dry wood. Choosing a short-backbone polymer, which is also a low viscosity polymer, will accomplish two things. The low viscosity of the solution will enable deep and thorough penetration of the wood; and once polymer-

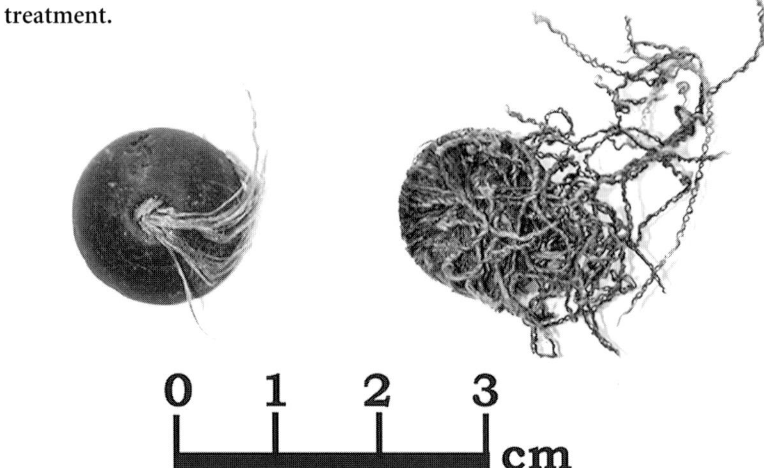

Fig. 3.6. Buttons after completion of the Passivation Polymer treatment.

ized, the smaller polymer will ensure that the wood is stronger than in its pretreatment state.

For dry cave-site wood, PR-10 with a 5% addition of CR-20 cross-linker, added by weight, is a perfect combination of polymers for treatment. Other polymer choices are listed in table 1.3.

1. Pour sufficient PR-10 into a container. To this solution, add 5% CR-20 cross-linker by weight and mix thoroughly. Always use a nonporous stir rod when mixing polymer solutions. This will ensure that equal proportions and stoichiometry are maintained. Unless thoroughly mixed, the less viscous material, in this case CR-20, could be absorbed by a porous stir rod, resulting in an improperly prepared polymer solution.

2. Carefully place the artifact into the PR-10/CR-20 solution. Regardless of which viscosity polymer has been chosen, be careful when immersing an organic substance. Delicate artifacts can be damaged if forced recklessly into a viscous bulking agent. Place a piece of aluminum screen over the artifact, making sure that the screen friction-fits against the sides of the container. This will ensure that the artifact remains immersed in the polymer solution throughout treatment. (See fig. 3.1.)

3. Dry wood needs little help to absorb a low viscosity polymer solution. If the conservator feels the need for pressure treatment, proceed with caution. Very little vacuum will be required. During vacuum-assisted impregnation, use as little vacuum as possible. As long as small bubbles flow slowly from the wood, there is sufficient pressure to ensure that the polymer will penetrate the artifact. If bubbling is too vigorous, reduce the vacuum pressure immediately. Excessive bubbling may damage delicate artifacts. Once bubbling appears to slow in frequency, or stop, increase the vacuum pressure and maintain pressure at a point where stream-type bubbling occurs. Experience and good observations dictate the amount of vacuum pressure that can be applied. If the diagnostic attributes of the artifact are not being compromised, it can remain in reduced pressure treatment for several hours. Because of their porosity, dried wooden artifacts averaging in thickness up to 1.5 inches are well penetrated after this type of processing.

4. If vacuum-assisted impregnation has been used, always make sure the artifact is returned to ambient pressure slowly. The artifact should then be taken out of the vacuum chamber and, after careful removal from the Passivation Polymers solution, placed onto an aluminum screen so that free-flowing materials can be recovered. There is no hurry to continue treatment on desiccated wood. The artifact can safely drain for several hours and, if necessary, be stored indefinitely in solution until continued treatment can occur.

5. Before a catalyst is applied, excess bulking agent needs to be removed from the artifact's surfaces. While many of the polymer solutions are sticky to the touch, they are easily wiped from the wood surfaces using Q-tips or clean cloths. To remove excess amounts of bulking agent from cracks and crevices, the artifact can be wiped with a rag soaked in CR-20. For complex surface-cleaning, or when all surface traces of bulking solution need to be removed, the artifact can be immersed in a bath of CR-20 then wiped with Q-tips and cloths.

6. After the artifact has been wiped, a suitable catalyst should be applied to all surfaces as soon as possible. A list of catalysts is supplied in table 1.2. Catalyst's attributes and application methods are also provided in that table. CT-32 is the best general-purpose catalyst for treating dried wood. After applying CT-32 and allowing it to sit on the surfaces of the wood for no more than two minutes, wipe thoroughly and quickly. Using the same procedures as noted for waterlogged wood, continue vapor deposition by placing the artifact into a Ziploc bag. Place a small

dish containing a few grams of CT-32 into the bag and seal it shut. The catalyzation process will continue for up to 24 hours.

If the surfaces of the wood feel damp, replace the used catalyst with a dish containing fresh catalyst. CT-32 has a maximum working time of 24 hours. Once the artifact feels dry, remove it from the Ziploc bag and allow it to cure in fresh air for as long as possible. While the artifact can remain in catalyzation as long as desired, the final stage of fresh air curing will allow final curing and ensure that catalyst fumes dissipate. If the wood requires spot treatment to remove white particulate formed on the surfaces during catalyzation, use a soft cloth containing a few drops of CR-20.

Reprocessing and Stabilization of PEG-Treated Wood

Re-treatment strategies have been of particular interest to conservators at the Archaeological Preservation Research Laboratory. For several decades, preservation of waterlogged wood using polyethylene glycol has been the standard treatment preferred by most conservators. A carbowax that is miscible in water, PEG is available in an almost infinite variety of molecular weights. A skillful conservator can blend two or more molecular weights of the bulking agent to attain specific and desirable traits for organic artifact preservation. Like all conservation strategies, PEG treatments are always being refined; recent advances in archaeological chemistry have expanded our knowledge of PEG chemistry and its applications in archaeology.

Re-treatment of PEG-treated artifacts is a concept that has grown out of necessity. When preserved using standardized treatment strategies developed by Per Hoffmann, David Grattan, and others in the 1980s and 1990s, artifacts treated with PEG are stable and well preserved. Artifacts treated in the mid-twentieth century, when development of conservation strategies was still in its infancy, have not fared as well.

PEG was used in the treatment of wood, leather, paper, and other organic materials, and was added to the water in which the artifacts were stored. Over time, the percentage of PEG was increased and the aqueous solution permeated into the cell structure of the artifact, depositing the carbowax bulking agent, which prevented cellular collapse. Because this preservation process was still in its developmental stages, conservators tended to use differing percentages of PEG in treatment, and in many cases, variance in post-treatment handling and curation of artifacts was substantial. Indeed, the pioneers of PEG conservation used freeze-drying in conjunction with PEG treatments. This combined treatment strategy yielded superior finished artifacts that were more stable and aesthetically correct.

But what about the hundreds of thousands of artifacts that were conserved without the benefit of advancements such as freeze-drying and blended PEG treatment strategies? Some artifacts have been lost. The majority of artifacts, however, have been carefully curated and re-treated over time. Early in 1992 Ole Crumlin-Pedersen of the Danish Viking Ship Museum traveled to Texas A&M University to discuss the vessels in the museum that had been preserved using PEG. He believed, as do I, that vessels preserved using traditional PEG treatment methods require additional treatment to make them as stable as possible. At Crumlin-Pedersen's encouragement, re-treatment of PEG-treated wood became a new challenge for conservators at Texas A&M University. The purpose of early experimentation was not to discredit the value of PEG in the preservation of waterlogged wood. On the contrary, research initiatives were directed at improving and strengthening the knowledge base for using PEG in archaeological preservation. While PEG is by far the most widely used bulking agent for the preservation of waterlogged wood, its basic characteristics offer some challenges for the long-term well-being of preserved wood.

The posttreatment hygroscopic nature of PEG is a concern when the carbowax compound is used for the long-term curation of structurally damaged, waterlogged wooden artifacts. When introduced into waterlogged wood, some of the PEG reacts with polysaccharides, polyphenolics, and water to form complex bonds that are hydrophobic in nature. The unbound PEG, located within the open structures of the wood, however, remains hydrophilic. Accordingly, fluctuations in temperature and humidity result in the uptake of atmospheric moisture and intercellular migration of the unbound PEG. In extreme conditions, the intercellular movement of PEG stresses the weak walls of the wood, causing shrinkage, distortion, and loss of diagnostic artifact attributes.

Damage caused by waterlogging is seldom uniform throughout a timber. Often, the inner core remains in excellent condition while its outer surfaces become soft and severely damaged. PEG treatments successfully displace water in the cell structure of waterlogged wood, but the bulking agent alone is usually not capable of stabilizing and protecting the surfaces. To stabilize badly deteriorated or surface-checked wood, an application of higher molecular weight PEG may act to consolidate the surfaces of the artifact from abrasive damage caused by dust and other airborne particles in the museum environment.

At the conservation facilities of Parks Canada in Ottawa, Canada, 834-403 Jade adhesive, a diluted polyvinyl acetate (PVA) emulsion, is used to consolidate the checked surfaces of PEG-treated wood.[1] The substance also strengthens the surfaces of PEG-treated timbers.[2] To ensure that badly deteriorated wooden artifacts have resilient outer surfaces, conservators at the Archaeological Preservation Research Laboratory at Texas A&M University have immersed PEG-treated wood in a saturated solution of acetone rosin, which achieves a result similar to PVA emulsion.

In the "Tongue Depressor Experiment," described below, CR-20, a hydrolyzable, multifunctional alkoxysilane was used to expedite the removal of unbound PEG from the wood. CR-20 self-condenses in the aqueous environment of the wood, forming a methyl silane polymer that is less susceptible to water absorption. Results of this experiment have shown that while PEG is an excellent bulking agent for archaeological preservation, it is not a reversible process. When unbound PEG is removed from wood in the presence of an alkoxysilane solution, the remaining PEG is hydrophobic and less prone to dimensional shifting as the result of environmental factors. The use of an alkoxysilane polymer to reprocess PEG-treated wood offers many advantages for the stabilization of waterlogged wooden artifacts. The process reduces, and sometimes eliminates, swelling that has occurred from introducing PEG into the waterlogged wood. After treatment, the surfaces of the wood are stabilized and do not require additional consolidation. All the wood samples regained a natural wood texture and color, and in all cases, the weights of the reprocessed samples were reduced.

Tongue Depressor Experiment

Nuclear magnetic resonance (NMR) analysis was used to investigate the chemical transformations that occurred when PEG 3350 was introduced into samples of waterlogged wood and to characterize chemical reactions resulting from the reprocessing of this wood using CR-20. During reprocessing in a vat of CR-20, unbound, free-flowing PEG was removed from the wood. In the aqueous environment of the PEG-treated wood, CR-20 displaced free-flowing PEG within the cell voids of the wood, hydrolyzing to form methyl silane polymers. Evaluation of the reprocessed wood indicated that swelling normally associated with traditional PEG preservation was eliminated. In many cases, samples of wood were returned to their pretreatment dimensions. Unlike PEG-treated wood samples, which became darker in color and waxy in surface appearance, the reprocessed wood appeared natural in coloration and surface texture.

Materials

Tongue depressors were chosen for experimentation because they are plentiful and uniform in dimensions, color, genus, and species. Several control tongue depressors were randomly collected to assist in evaluating PEG-treated wood. To create a supply of waterlogged wood, hundreds of white birch *(Betula papyrifera)* tongue depressors were placed into a four-liter stainless steel container and boiled in freshwater for 10 days. Water was added at regular intervals to ensure that the water level remained high enough to keep all the wood pieces saturated.

The tongue depressors were then stored at room temperature in a large jar of water with a tight-fitting screw cap. One tongue depressor, labeled TD2, was stored separately in freshwater for comparison with the control sample, TD1, and the PEG re-treated wood, TD3. Several pieces of wood were allowed to air-dry at room temperature for 24 hours. After air-drying, all samples exhibited extensive warping and shrinking.

At room temperature, PEG 3350 is a powder. To make the task of incremental additions easier, a large jar of PEG 3350 was placed into a warming oven and warmed to a temperature of 30°C. At this temperature, it remains in a liquid state. CR-20 is a clear monomer that reacts with water to form silanetriol and methanol. The silane condenses with available hydroxyl groups or other silanol monomers to form siloxane resins. CT-32 is a tin- based catalyst that remains active for approximately 24 hours in a controlled environment. It can be applied topically to the surfaces of the silicone oil–impregnated artifact or as a vapor. Vapor deposition accelerates the chemical reactivity of the catalyst and ensures an even application to all surfaces of the artifact.

Other materials included a Pyrex graduated cylinder (250 ml), a piece of aluminum foil to form a loose-fitting cap for the graduated cylinder, and a Despatch LFD series warming oven with computerized temperature and humidity controls.

Methodology

Twenty waterlogged tongue depressors were randomly selected, thoroughly rinsed in running tap water, and stored in a large jar of freshwater. The jar was then placed into a vented warming oven set at a constant temperature of 70°C. Small amounts of water were added to the container each day to maintain a constant level throughout treatment. Each week, one 10% increment of warmed PEG 3350 (by volume) was added, forming an aqueous solution of PEG. After seven weeks, the warming oven was turned off; the tongue depressors were allowed to sit in the PEG solution for another week.

One tongue depressor, labeled TD3, was removed from the PEG solution, placed between paper towels to blot free-flowing PEG from its surfaces, and allowed to sit in fresh air for 24 hours before additional experimentation was conducted. The length, width, thickness, and weight of TD3 were then recorded, and NMR analysis was performed. (The waterlogged sample, TD2, and the untreated sample, TD1, were also subjected to NMR analysis.) TD3 was then immersed in 200 ml of CR-20 in a graduated cylinder. To prevent rapid evaporation of the silane, the graduated cylinder was loosely capped with a three-inch square of aluminum foil. The cylinder was placed into a 70°C vented warming oven for 24 hours (fig. 3.7). The cylinder was then removed from the oven and allowed to cool to room temperature and sit for an additional 5 hours. When TD3 was removed from the CR-20 solution, it was covered with a thin white coating of PEG. This was removed by wiping the tongue depressor with a lint- free cloth. The remaining CR-20 solution in the graduated cylinder was milky white due to PEG extracted from the tongue depressor. This solution was poured into a flat pan and allowed to sit in a fume hood for several days.

Vapor deposition was chosen as the best method to ensure that all surfaces of the wood were evenly exposed to catalyst. To form an effective environment in which warmed vapor fumes would remain in close contact with the surfaces of TD3, a tall polyethylene container with a tight-fitting lid was used to create a containment chamber. In an inverted position, the lid of the container became the base of the unit.

A Fisherbrand aluminum weighing dish with 15 g of DBTDA catalyst was placed in the middle of the base of the chamber. A piece of aluminum screen was placed over the dish to act as a platform on which the tongue depressor could be placed during the catalyzation process. Paper towels were placed on top of the aluminum screen to prevent free-flowing PEG/CR-20 solution from contaminating the catalyst during treatment (fig. 3.8). When the body of the chamber was snapped into place over the tight-fitting base, an ideal environment for vapor deposition of the catalyst was created. TD3 was left in catalyst vapor deposition for 24 hours in a warming oven set at 70°C. It was then removed from the oven and exposed to fresh air in a fume hood. After 24 hours in air, TD3 was evaluated and samples of wood were collected for NMR analysis.

NMR spectrographic analysis showed that CR-20 was not polymerizing PEG within the cell voids of the wood, as had been expected. Instead, the warm solution of CR-20 was removing the unbound PEG solution from the cells. PEG remains water miscible within the cells of wood. It, therefore, remains sensitive to temperatures. When immersed in warm CR-20, PEG melts. Because it is not chemically bound to the cell walls of the wood, it migrates. This is the same reaction that would be observed if the PEG-treated wood were placed into warm water.

Of interest, then, is the chemical reactivity of CR-20 within the cell structure of the wood. PEG was introduced into the wood as an aqueous solution. Accordingly, water was bound to the PEG. It appears that H_2O asso-

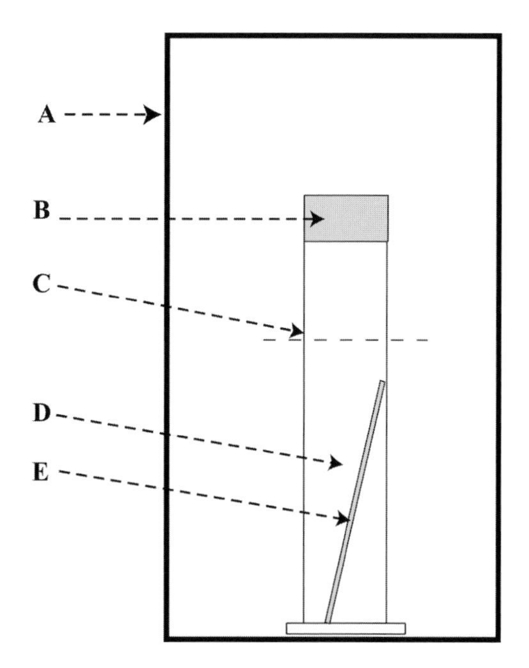

Fig. 3.7. Graduated cylinder setup for reprocessing of PEG-treated wood using CR-20: *(A)* vented warming oven; *(B)* loose-fitting aluminum cap; *(C)* level of CR-20 in cylinder; *(D)* CR-20 solution in which TD3 was immersed; *(E)* TD3 immersed in CR-20.

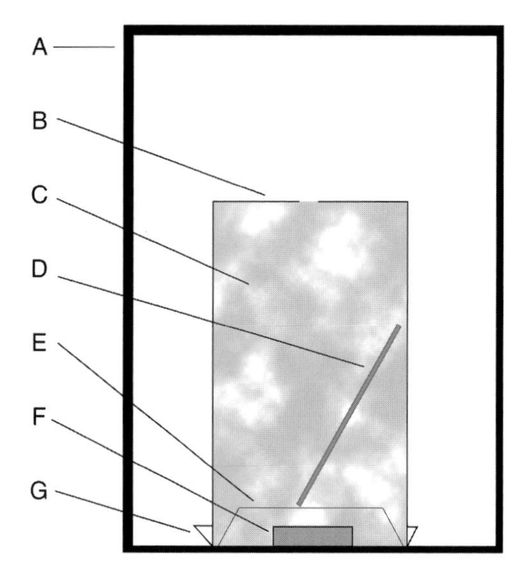

Fig. 3.8. Configuration of the containment chamber used for vapor deposition: *(A)* warming oven; *(B)* upper body of polyethylene chamber, with a small hole in its top; *(C)* CT-32 vapor created by warming the catalyst to 70_C; *(D)* TD3 sitting on top of the source of catalyst; *(E)* aluminum screen and paper towel, creating the platform on which TD3 sits; *(F)* aluminum weighing dish holding 15 g of CT-32 catalyst; *(G)* tight-fitting base of the containment chamber.

ciated with PEG had reacted with CR-20, acting as part of the self-condensing reaction that had formed new polymers within the cell structure of the wooden tongue depressor. Further NMR analysis was conducted to understand this chemical reaction.

^{1}H, ^{13}C, and ^{29}Si Nuclear Magnetic Resonance Characterization

The solid state NMR spectra were acquired with a Bruker MSL 300 spectrometer operating at a magnetic field of 7.05 Tesla. The high resolution solid state ^{1}H (CRAMPS) spectra were acquired with an MREV-8 pulse sequence using a 2 microsecond 90-degree pulse and a tau value of 4 microseconds. The ^{1}H-^{13}C cross-polarization (CP/MAS) spectra with high power 1H decoupling and magic angle spinning were acquired with a 90-degree pulse of 4 microseconds, a contact time of 1.5 milliseconds, and a recycle delay of 5 seconds.

Additionally, ^{13}C spectra (Bloch decay) were acquired with a 45-degree excitation pulse with high power ^{1}H decoupling and magic angle spinning using a recycle delay of 10 seconds. The ^{29}Si spectra (Bloch decay) were acquired using a 45-degree excitation pulse with high power ^{1}H decoupling and magic angle spinning using a recycle delay of 20 seconds.

The ^{13}C CP/MAS spectrum for the untreated control tongue depressor (TD1) is shown at the bottom of figure 3.9. The spectrum is consistent with a composition of cellulose, hemicellulose, and lignin.[3] Both cellulose and hemicellulose are carbohydrates. The ^{13}C resonances of cellulose are between 60 and 110 ppm. Hemicellulose contains acetate functional groups with resonances at 22 and 174 ppm. Lignin is a polymer of phenolic cinnamyl alcohols. No significant spectral differences were noted in comparison with the Bloch decay spectrum. The ^{13}C CP/MAS spectrum for the waterlogged tongue depressor (TD2) is shown at the top of figure 3.9. The preparation of TD2 by boiling in water has been shown to alter the chemical constitution of the tongue depressor, as evidenced by the loss of the acetate resonances at 22 and 174 ppm in comparison with the spectrum on the control (TD1). The changes in these ^{13}C spectra of the two birch wood tongue depressors are quite similar to those reported by Wilson et al. for oak woods taken from shipwrecks.[4] The ^{13}C spectral signature, along with the macroscopic observations of extensive warping and shrinking upon drying in air, suggest that this preparation provides a suitable model for the analysis of waterlogged wood.

The ^{13}C CP/MAS spectrum of the waterlogged tongue depressor (fig. 3.9, *top*) is nearly identical to that of the waterlogged tongue depressor (TD3) after treatment with PEG 3350 (fig. 3.10, *bottom*). The spectral difference for the waterlogged tongue depressor treated with PEG 3350 shows up in the ^{13}C Bloch decay spectrum (fig. 3.10, *top*). A large resonance at 70.6 ppm is present, in good agreement with the sharp single resonance of 70.2 ppm observed in the ^{13}C solution state spectrum for an aqueous solution of PEG 3350 and with the broader ^{13}C resonance at 71 ppm for PEG 3350 in the solid state.

The absence of signal or small signal intensity observed under cross-polarization is consistent with significant molecular mobility of the PEG 3350. Such mobility would attenuate or average the ^{1}H-^{13}C dipolar interaction used for magnetization transfer in the experiment. This observation of the ^{13}C resonance of the PEG 3350 in the Bloch decay spectrum and not in the cross-polarization spectrum indicates that the PEG 3350 is still a very molecularly mobile species, even when absorbed within the wood. While molecular mobility sufficient to attenuate the dipolar interaction necessary for cross-polarization need not include translational mobility, only slight temperature increases above ambient are necessary for the PEG 3350 to migrate out of the wood.

TD3 was then treated with CR-20. The ^{13}C CP/MAS spectrum of the waterlogged tongue depressor treated with both PEG 3350 and CR-20 (fig. 3.11, *bottom*) is nearly identical with that of the waterlogged tongue depressor (TD2) shown in figure 3.9 *(top)*. The difference is a new ^{13}C resonance, which appears at -3.2 ppm. It should be noted that the ^{13}C Bloch decay spectrum (not shown) of this sample still indicates the presence of some PEG 3350. Inside the tongue depressor, a condensation reaction has occurred. The CR-20 appears to have been hydrolyzed in the aqueous environment to the triol, giving off methanol. The triol then self-condensed to form a silicone polymer. The ^{13}C CP/MAS spectrum of the solid polymer formed under identical conditions in the absence of the wood is shown in the top of figure 3.11. The major resonance is at -3.1 ppm with small resonances at 49.6 and 70.5 ppm. In aqueous solution, the ^{13}C resonance of PEG is at 70.2 ppm.

Though CR-20 is not miscible in aqueous solution, the ^{13}C solution state spectrum acquired in deuterated chloroform shows a methoxy ($-SiOCH_3$) resonance at 49.6 ppm and a methyl ($-SiCH_3$) resonance at −9.5 ppm. This indicates that only small amounts of PEG and methoxy groups are present with a change in the methyl group resonance as the polymer is formed.

No ^{29}Si resonances were observed for the control tongue depressor (TD1). However, figure 3.12 shows the ^{29}Si spectra of the waterlogged tongue depressor (TD3) treated with both PEG 3350 and CR-20 *(bottom)* and of the solid polymer formed from the CR-20 *(top)*. The spectra are consistent with two types of silicon environments, one with silicons bearing two siloxy $\{H_3CSi(-OSiO)_2OR\}$ groups (−56 ppm) and one with silicons bearing three siloxy groups (−65 ppm).

Since the ^{13}C spectrum of the solid polymer formed from the CR-20 (fig. 3.11, *top*) showed only a trace of a methoxy resonance, it is clear that the methoxy is not the fourth bond on the silicon sites with a methyl and two siloxy groups. The presence of hydroxyl groups as this fourth bond is confirmed in the solid state ^{1}H spectrum of the polymer (fig. 3.13). In this spectrum, the peak at -0.2 ppm shows the methyl protons on the silicon while the smaller peak at 3.1 shows the hydroxyl protons on the silicon. The structure

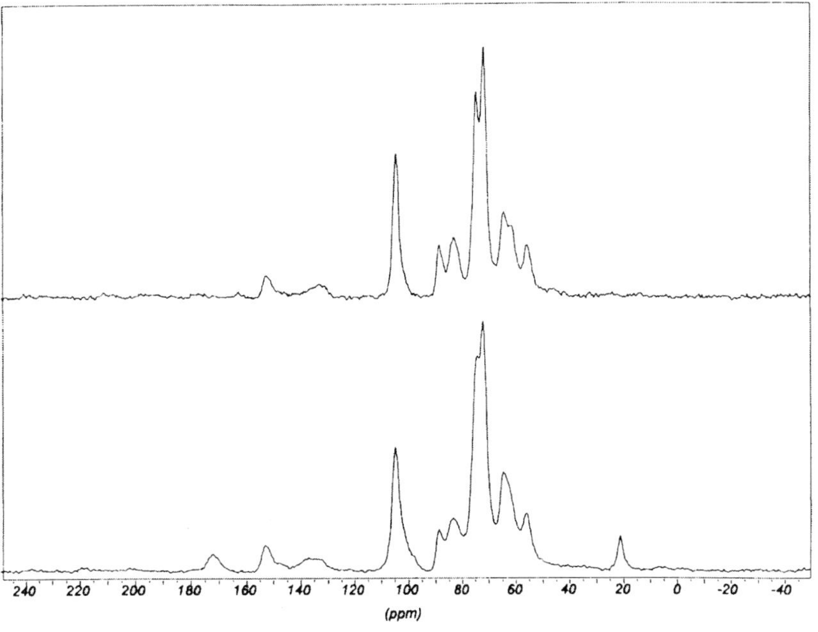

Fig. 3.9. ^{13}C CP/MAS spectra of the control tongue depressor TD1 *(bottom)* and the waterlogged tongue depressor TD2 *(top)*.

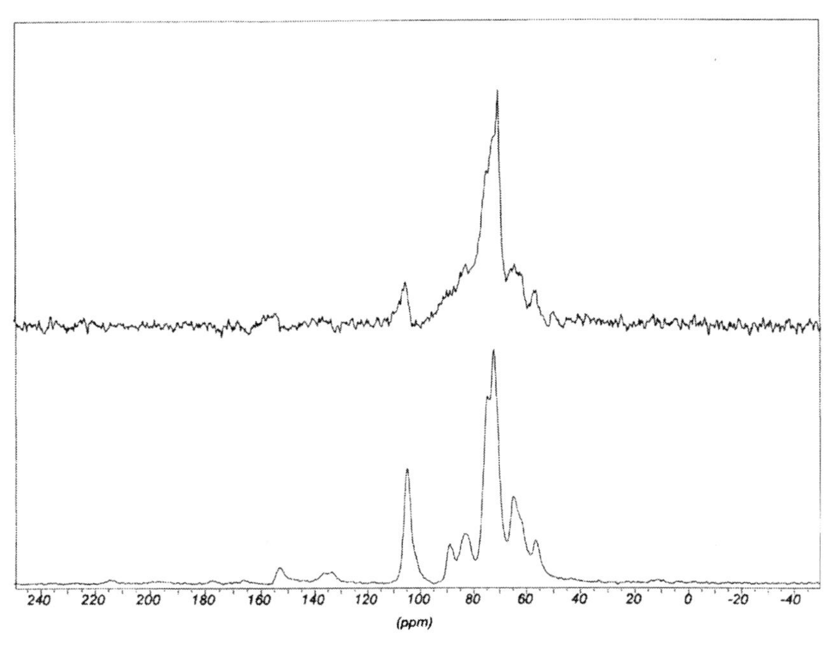

Fig. 3.10. ^{13}C CP/MAS *(bottom)* and Bloch decay *(top)* spectra of the waterlogged tongue depressor TD3 treated with PEG 3350.

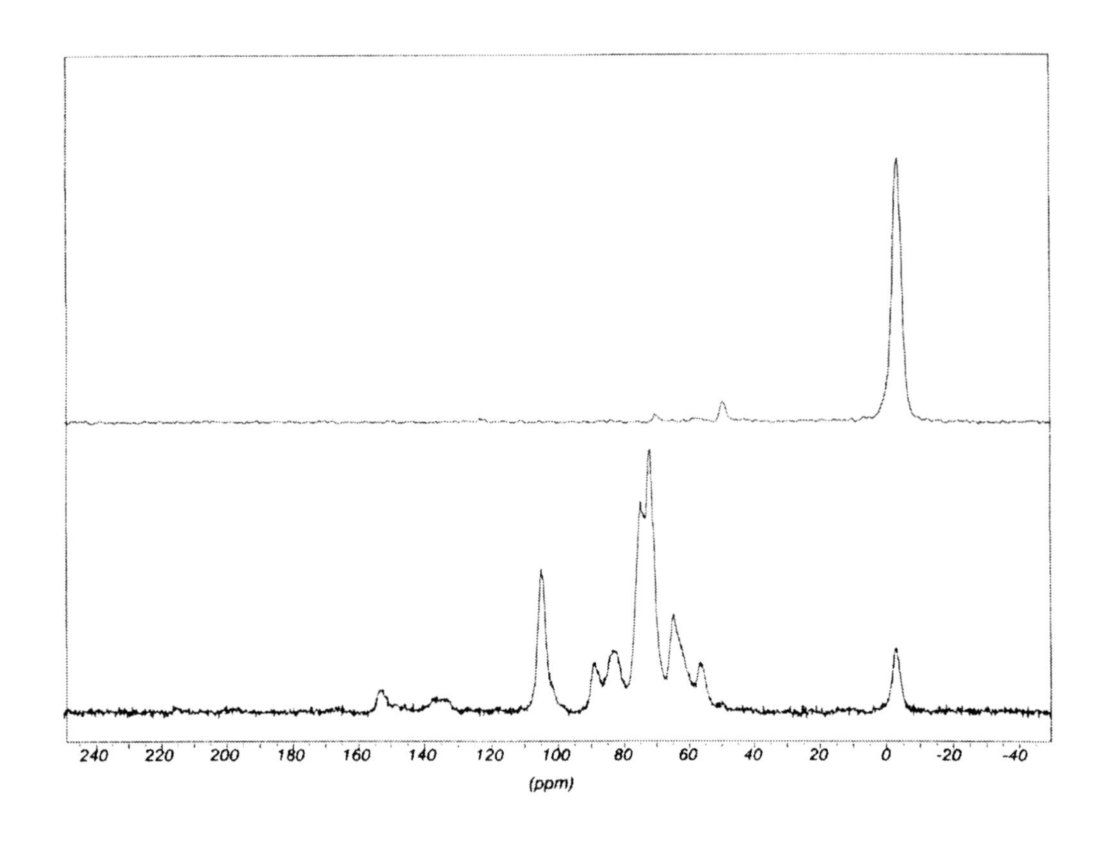

Fig. 3.11. ^{13}C CP/MAS spectra of the waterlogged tongue depressor TD3 treated with both PEG 3350 and CR-20 *(bottom)* and of the solid polymer formed from the CR-20 *(top)*.

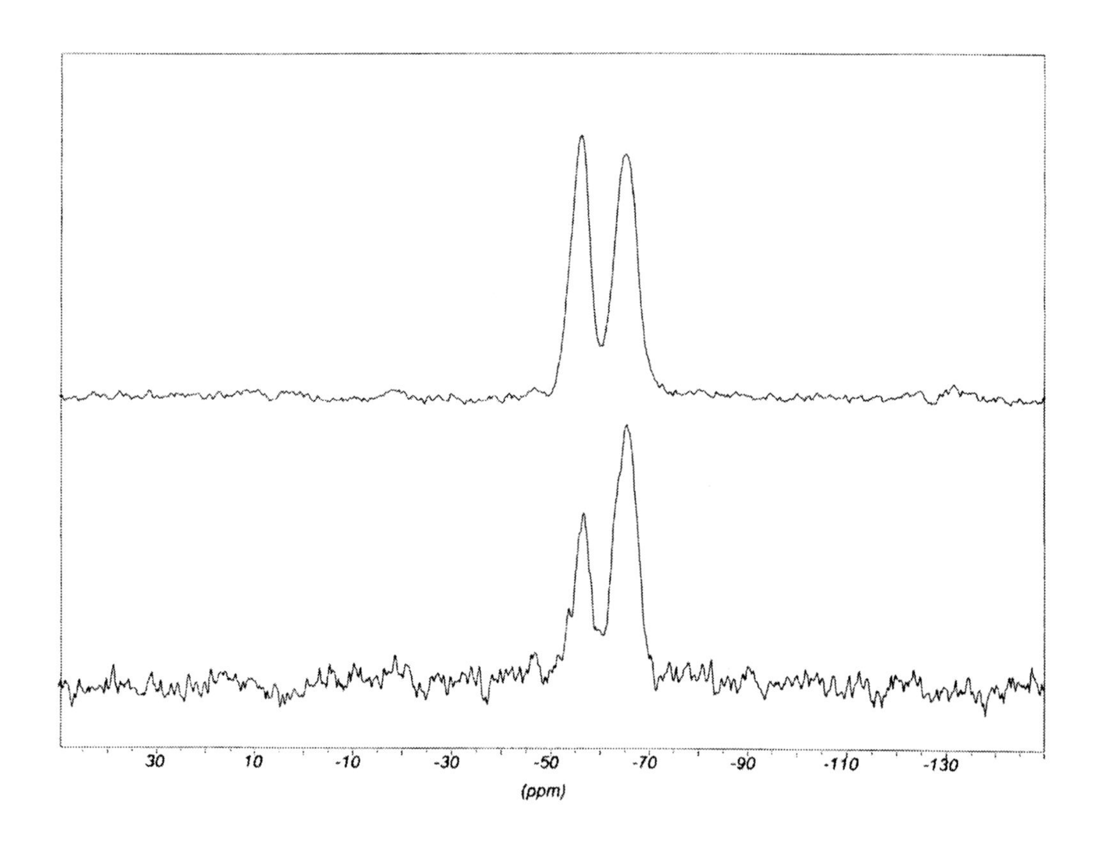

Fig. 3.12. ^{29}Si/MAS spectra of the waterlogged tongue depressor TD3 treated with both PEG 3350 and CR-20 *(bottom)* and of the solid polymer formed from the CR-20 *(top)*.

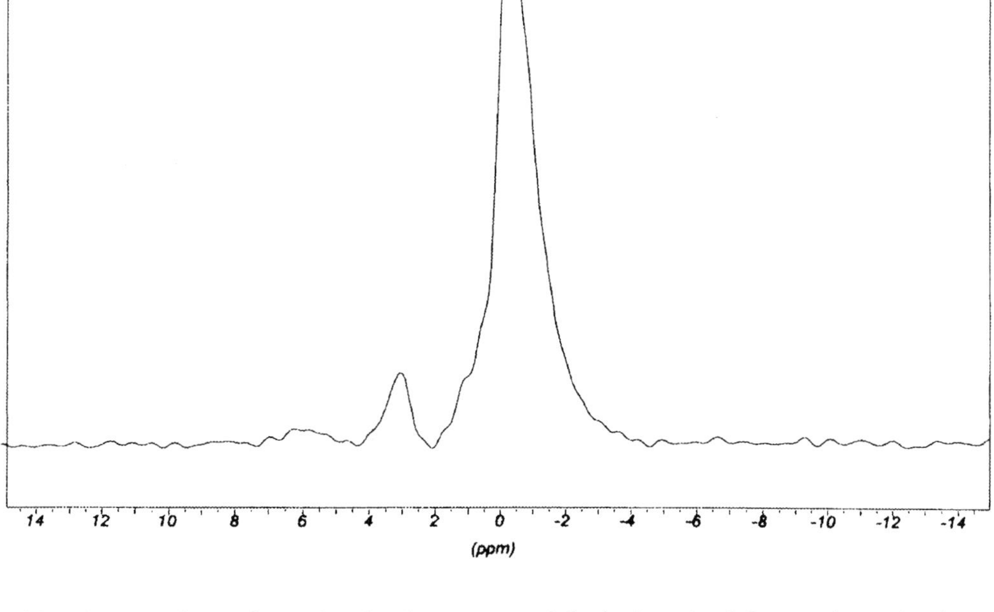

Fig. 3.13. ¹H CRAMPS spectrum of the solid polymer formed from the CR-20.

of the silicone polymer formed in the absence of the wood as well as within the wood is shown in figure 3.14.

Observations

The PEG-treated tongue depressor TD3 was a translucent, strong brown color (Munsell 7.5YR 4.6), and the surface texture of the wood was waxy and slightly sticky. After treatment in CR-20, however, the wood changed to a gray brown color (Munsell 10YR 8.2). The presence of heat and moisture during immersion in CR-20 encouraged the flow of PEG 3350 from the cells of the wood so that, after the treatment, the surfaces of TD3 felt dry. Additionally, the surface texture and wood grain of TD3 were natural in appearance, even though waterlogging had changed the color.

Measurements of the control tongue depressor (TD1) were used to monitor dimensional changes for TD3 at various stages of treatment. TD1 measured 15.269 cm long and 1.832 cm wide at the center. After preservation in PEG 3350, swelling was noted in TD3. While the length of the wood remained unchanged at 15.249 cm, its center width swelled to 2.009 cm, a 9.66% increase. After removal of PEG from the wood and treatment in CR-20, swelling was almost completely eliminated from TD3. The length of TD3 remained constant at 15.249 cm, and the width was reduced to 1.830 cm, a decrease of 8.91%. The weight of TD1, the control tongue depressor, was 2.8 g. After treatment in PEG, TD3 weighed 6.1 g, a 217.9% increase. After

CH_3 CH_3

| |

- O - Si - O - Si - O -

| |

O O

Fig. 3.14. Structure of the silicone polymer.

treatment in CR-20, the weight of TD3 was 3.4 g, a 44.26% decrease.

Prior to treatment, temperature variations affected TD3. At slightly elevated temperatures of 28°C, the surfaces of the wood were sticky; at 40°C, its surfaces were wet. After treatment in CR-20, however, stickiness and pooling of PEG were not noted when TD3 was exposed to elevated temperatures. The CR-20-treated wood remained dry without signs of warping.

The surfaces of TD3 were slick with a thin layer of PEG when the wood was removed from the CR-20 solution. After wiping with lint-free cloths, the wood looked more like the original color and texture of the control wood sample. After a few minutes of exposure to fresh air, the surfaces of the wood were dry, without additional color changes. When the CR-20 was cooled to room temperature, a suspension of PEG was noted in the solution (fig. 3.15). This solution was poured into a shallow pan and allowed to sit in a fume hood for several days. Once exposed to fresh air, the solution formed a hard, clear polymer.

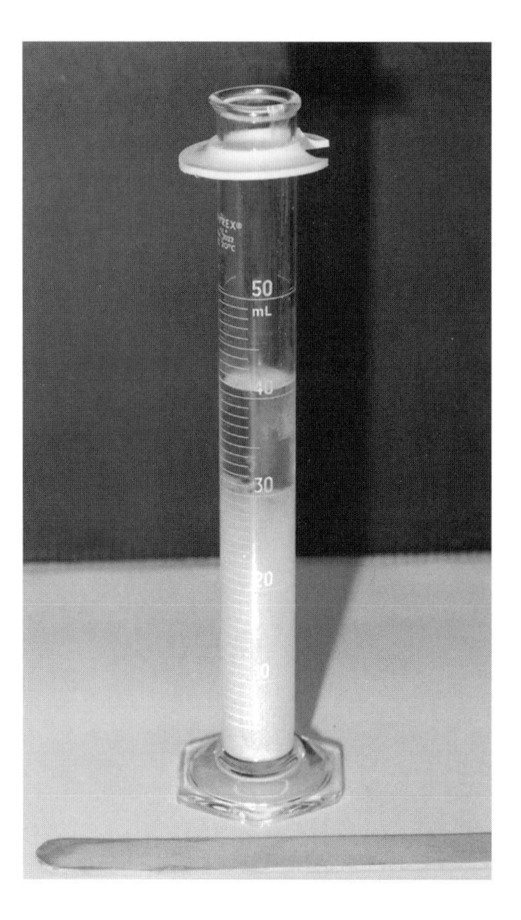

Fig. 3.15. PEG from TD3 suspended in CR-20 solution.

The solution of CR-20, PEG, and water that resulted from the extraction of free-flowing PEG from wood formed hard polymers when the solution was exposed to fresh air for three days. There was some ^{13}C, NMR evidence to indicate that the CR-20 also reacted with some of the hydroxyl end groups of the PEG, aiding in the formation of new polymers within the cell wall structures of the wood. Scanning electron microscopy was conducted on samples of wood prior to and after treatment in CR-20. In figure 3.16, the cell structure of the wood is not evident due to the heavy concentration of PEG. After treatment in warmed CR-20, however, PEG that had not bonded to cell walls was eliminated. At higher magnification, the bond between PEG and interior cell walls is evident (fig. 3.17).

Conclusions

Silicone polymers are formed when PEG-treated wood is re-treated with CR-20 monomers. These polymers are hydrophobic in nature, and their presence appears to make the wood less sensitive to atmospheric moisture. This is a potential benefit for the treatment of composite artifacts when the stability of closely associated metal components is an issue. Silicone oils can be used for spot treating organic materials. The silicone oil/CR-20 solution can be topically applied using Q-tips or small brushes. After allowing the solution to permeate into the organic substrate, a catalyst can be applied topically or as a vapor. This allows the conservator to specially treat difficult areas of a PEG-treated artifact or areas that may be in contact with metals.

Little is known about the various byproducts created from the decomposition of PEG over time and the potential contaminants resulting from long-term deterioration.

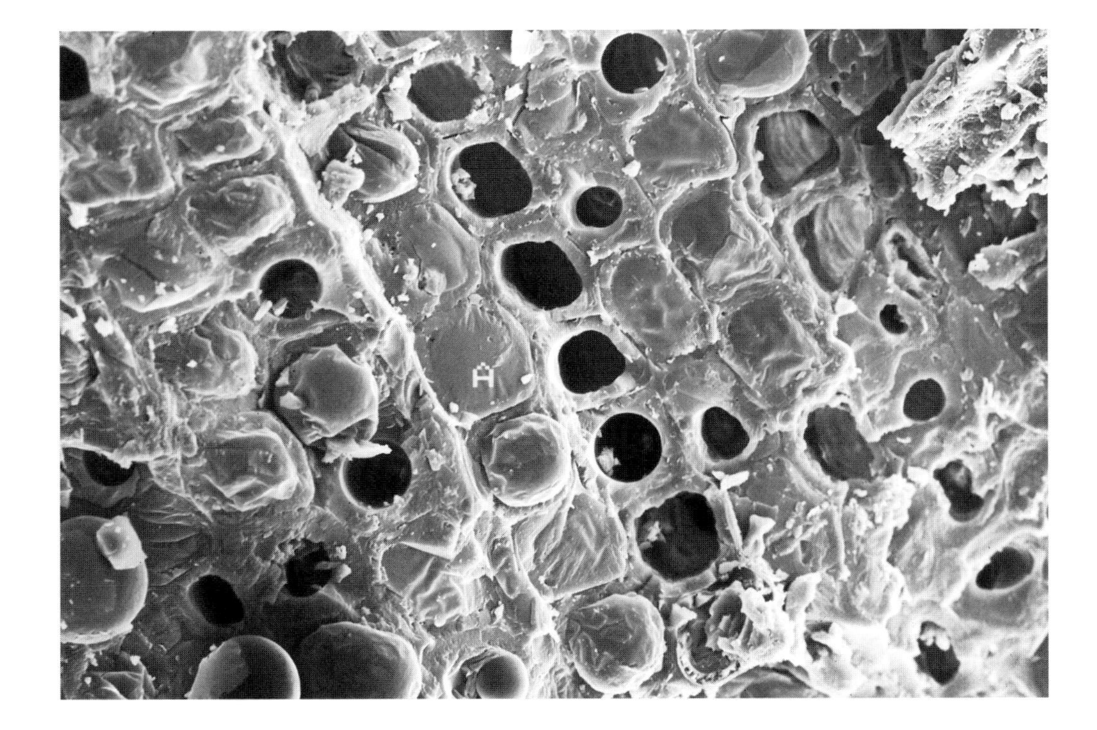

Fig. 3.16. PEG 3350 in waterlogged wood.

One potential by-product is formic acid.[5] Polymers formed in solution with CR-20 are more rigid in structure, and in this state, formic acid is not formed. Methanol, however, is formed in the hydrolysis of CR-20. Subsequent testing has shown that catalyzation of PEG-impregnated artifacts is not necessary. CR-20 hydrolyzes in the aqueous environment to the triol. It then self-condenses to form the silicone polymer, forming two distinct silicons illustrated in the ^{29}Si spectra (see fig. 3.12). The silicon environment has a methyl group and three siloxy bonds (i.e., So-O-Si). The second silicon environment has only two bonds. The third is formed to the methyl group while the fourth bond is to the hydroxy group as in the ^{1}H spectra (see fig. 3.13). Polymerization can be accomplished without catalyzation. The use of catalysts should not be ruled out, however. Additional experimentation is needed to determine if catalyzation acts to strengthen or enhance the polymers created by combining PEG with CR-20.

Repeated testing has shown that immersion in heated CR-20 removes unbound PEG from the cells of waterlogged wood, probably due to heat-induced migration. After treatment in CR-20, PEG bonded to cell walls is stable and resistant to environmental influences. One advantage to the CR-20 treatment process is that, after treatment, the diagnostic attributes and general aesthetics of the wooden artifact are well preserved. Due to the process of degradation caused by immer-

Fig. 3.17. After free-flowing PEG is removed from the wood, PEG bound to the cell walls is evident.

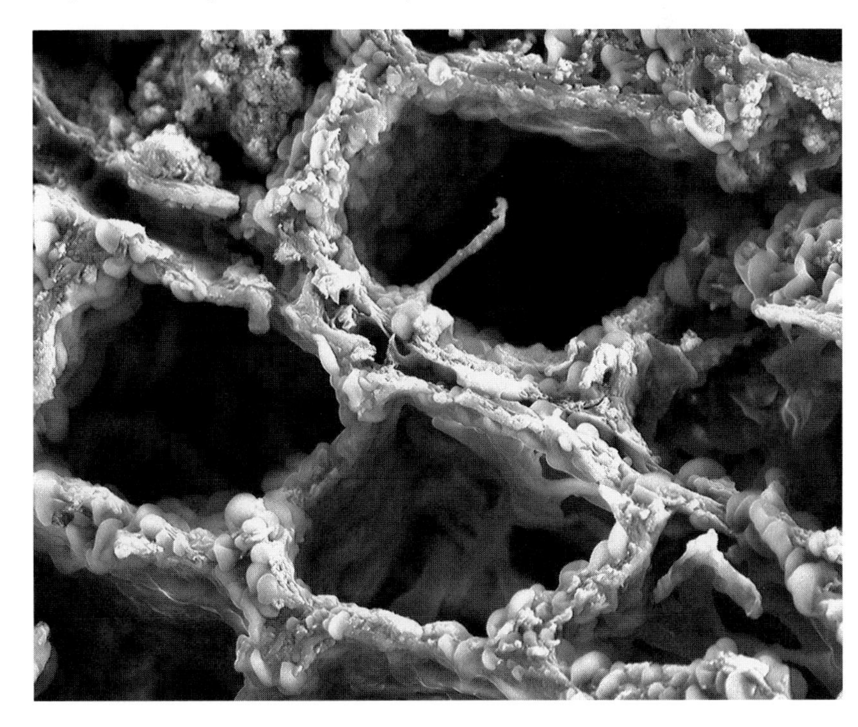

sion in water, CR-20- treated wood does not have the same coloration as the nonwaterlogged control wood samples. The wood, however, is dimensionally stable after treatment in the silane. Because unbound PEG has been removed from the cell structure, the wood is lighter in weight.

After several days' exposure to fresh air, the CR-20/PEG solution resulting from extraction of PEG from TD3 had formed into a hard, clear polymer. CR-20 self-condenses to form this polymer, with or without PEG. Attempts at dissolving the polymer in water, alcohol, or acetone have not been successful, and the polymer is not affected by fluctuations in temperature or humidity. Additional experimentation is needed to determine if the CR-20 polymer, which had bonded to the cell walls of the wood, strengthens the artifact. Preliminary results indicate this may be the case. Because unbound PEG flows freely from the artifact during treatment in CR-20, cell interiors and voids within the wood remain unclogged. Removal of free-flowing PEG results in a reduction in weight of the wooden artifact. This reduces physical stress on the wood. The resultant polymers do not fill cell structures and voids within the wood, making additional re-treatment possible, if needed.

Because of the self-condensing nature of CR-20, resulting in the formation of polymers, catalyzation is not necessary. In the case of the waterlogged tongue depressors, the use of topical and vapor deposition catalyzation does not appear to adversely affect the end results of CR-20 treatment.

Case Study: Re-treatment of Two PEG-Treated Sabots

Numerous books have been written on the history of the machinery of war, and exhaustive studies have investigated technological advances prior to and during the American Civil War. One interesting adaptation to spherical case ammunition often associated with large canons is the sabot, a wooden, flat-based attachment that allowed better positioning of the ammunition and fuse within the breech of a gun. In this case study, I examine an effective process developed at the Archaeological Preservation Research Laboratory to stabilize the wooden components of a composite artifact initially preserved using polyethylene glycol.

The two sabots delivered to the lab were part of an assemblage of 62 Civil War spherical shot cases recovered from Mirror Lake near North Calais, Vermont. These shot cases were intact when recovered, but the flat-based wooden sabots had been separated from the cast iron components for conservation purposes. First, each of the shells was considered live, necessitating that each be defused. Second, it was determined that the component parts required separate conservation and that each part would need close monitoring in wet storage. Figure 3.18 is a more modern example of an assembled spherical shot.

Inspection of the two wooden artifacts indicated that, even three years after treatment, both remained wet to the touch and surprisingly spongy. Because both sabots were very wet, it was impossible to section

Fig. 3.18. Sabot and shot.

Fig. 3.19. Sabots A and B.

and visually identify the genus and species of the wood used for construction. Accordingly, the decision was made to attempt this identification after the wood had been retreated. Sabot B was in better condition than sabot A, which had numerous cracks around the upper edge of its concave surfaces and a central hole and crack running through the base. In both cases, raised wooden grain along the tangential plane suggested that, at some point, the specimens had been partially dehydrated, causing shrinkage and slight distortion of the surface wood. As E. E. Astrup has noted, the problem of shrinkage is expected because of evaporation of residual water within a treated specimen and the resultant decrease in the volume of PEG.[6]

Conservation notes for the two artifacts indicated that both sabots had been mechanically cleaned with dental instruments. Each artifact was then rinsed in a 10% bath of hydrochloric acid to remove iron stains prior to rinsing in a 4% solution of hydrogen peroxide with 4% ammonia in deionized water to remove organic stains. After additional rinsing, the sabots were treated in a blend of 400 and 540 PEG, which was slowly increased to a 50/50 water/PEG mixture. Following treatment, the sabots were surface-wiped and then freeze-dried to remove excess moisture. Upon completion of the preservation process using PEG, it was evident that it would be impossible to reassemble the wooden bases with their conserved components, due to the hygroscopicity of the wood. Sabots A and B are illustrated in figure 3.19, and the exterior surface texture of the sabots is illustrated in figure 3.20.

Fig. 3.20. The exterior surface of one sabot.

Prior to re-treatment, each of the sabots was weighed, measured, and photographed to record all aspects of the artifacts for comparative analysis following re-treatment. Because neither sabot appeared to be perfectly round, two sets of diameter measurements were recorded. The position of the first measurement was recorded, and a second diameter reading was taken at 90 degrees to the first. Tracings of the top and bottom surfaces of the sabots were made to assist in monitoring potential dimensional changes that may have resulted from re-treatment. Dimensional data appear in table 3.2.

Test Process in Preparation for Treating the Sabots

An initial experiment was conducted to determine the efficacy of an alkyl silane for removing free-flowing PEG from the cellular structure of waterlogged wood. This type of cross-linker is commonly used in the Research Laboratory as a cross-linking agent for the purpose of polymerizing dimethyl siloxane, hydroxyl-terminated polymers. It has proved potentially useful for treating chemically bound PEG within the cell wall structures of wood, reducing the hygroscopicity of the remaining PEG.

To illustrate the general results from repeated experimentation, a single waterlogged tongue depressor (TD), treated with PEG 3350, was taken from its PEG solution and wiped with a paper towel to remove pooled PEG from all its surfaces. The sample was then placed into a graduated cylinder containing 50 ml of fresh CR-20. A loose-fitting aluminum foil cap was placed over the top of the cylinder. Then the TD, immersed in solution, was placed into a vented warming oven set at 71.1°C for 24 hours (fig. 3.21). The sample was then removed from the oven and allowed to stand at room temperature for 5 hours before it was taken out of the CR-20 and surface-wiped with a paper towel. Inspection of the cylinder revealed that a large volume of free-flowing PEG had been extracted from the tongue depressor. As the CR-20 solution cooled, PEG remained in suspension, appearing as a fluffy white substance.

The last step in the process of re-treating the PEG-treated TD was to initiate vapor catalyst deposition, using a tin-based catalyst to complete the polymerization process. To accomplish this, a closed environment was needed to contain catalyst fumes in close association with the tongue depressor. A containment chamber was constructed out of a one-quart polyethylene pail placed in an inverted position so that its tight-fitting cover served as the flat base of the unit. A catalyst tray was created using an aluminum sample tray with mesh aluminum screen placed over it. The screen formed a platform on which the sample rested and was positioned above the catalyst tray during catalyst vapor deposition processing. Several paper towels were placed on top of the screen to prevent free-flowing PEG from dripping into, and contaminating, the catalyst. Fifteen grams of CT-32 were added to the catalyst tray, and the TD was positioned on the screen with the body of the containment chamber properly fitted over it (fig. 3.22).

Once assembled, the unit was placed into

Table 3.2 Post–re-treatment Measurements for Sabots A and B

Sample	*Top Width (in cm)*	*Base Width of Rim (in cm)*	*Height (in cm)*	*Weight (in g)*
Sabot A	11.505 (1st reading)	9.72 (1st reading)	4.882	2.662
	11.026 (2d reading)	9.75 (2d reading)		
Sabot B	10.986 (1st reading)	10.00 (1st reading)	4.65	2.622
	10.984 (2d reading)	9.73 (2d reading)		

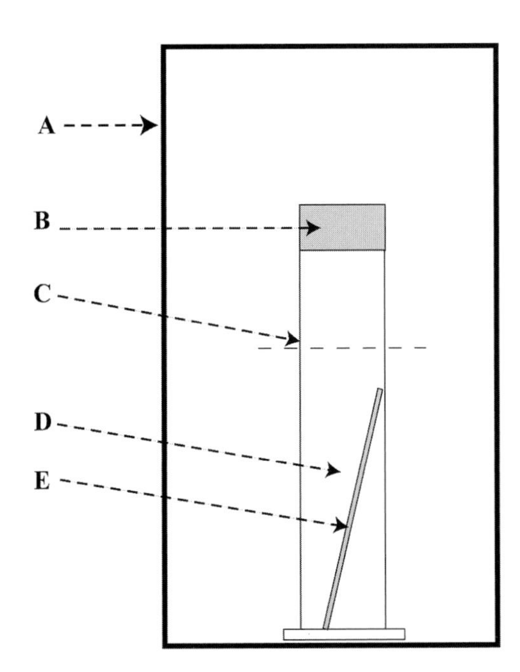

Fig. 3.21. Configuration of the containment chamber used for catalyst vapor deposition: *(A)* vented oven; *(B)* loose-fitting cap; *(C)* graduated cylinder; *(D)* CR-20 solution; *(E)* tongue depressor.

a vented warming oven set at 71°C. The sample remained in catalyst vapor deposition for 24 hours. Following treatment, the sample was removed from the oven and transported to a vented fume hood and opened. After the TD was allowed to sit in fresh air for several days, it was placed into a vented warming oven and heated to 71°F for a few hours. Throughout the heating process, no dimensional changes or sweating were noted.

Observations of the Test Process

After bulking with the PEG 3350 and subsequently being wiped with paper towels, the treated TD was dark brown. The edges of the TD were translucent, and its surfaces felt waxy and smooth. After re-treatment to remove free-flowing PEG 3350, the tongue depressor was removed from the oven, gently surface-wiped, and allowed to sit in fresh air. After returning to room temperature, the CR-20 had changed from a clear, watery solution into a milky colored solution as a result of the PEG in suspension, which had been removed from the tongue depressor. Instead of remaining dark brown, the tongue depressor was a light gray brown. The re-treatment color was very similar to the color of the control samples. Prior to re-treatment, the PEG-treated tongue depressor had swelled to a size slightly wider than the control tongue depressor. After re-treatment, the tongue depressor had returned to its original width.

Re-treatment of PEG-Treated Waterlogged Wood

The methods used to preserve sabots A and B were the same as those outlined for the tongue depressor above. Sabot B was placed into a large plastic container and immersed in a sufficient volume of CR-20 to ensure that there was approximately ½ inch of CR-20 covering the uppermost portion of the artifact. A loose-fitting top was positioned over the container, and the artifact in solution was placed into a well-vented warming oven set at 71°C. The artifact was left in treatment for 4.5 hours.

Sabot A had extensive cracks that extended through its sides and base. To determine if the presence of a catalyst in the cross-linker

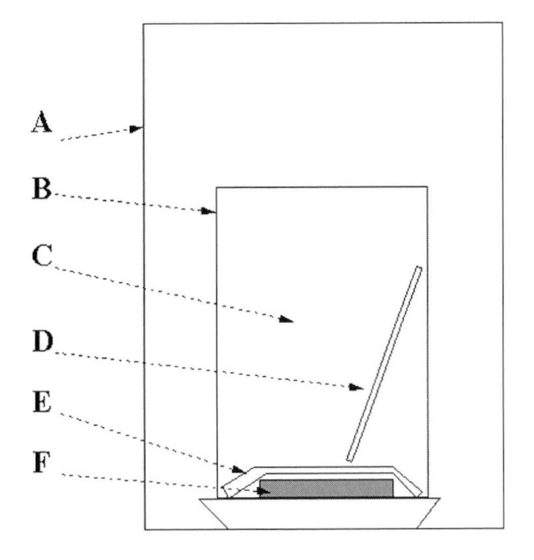

Fig. 3.22. Configuration of the containment chamber used for catalyst vapor deposition: *(A)* warming oven; *(B)* glass containment chamber; *(C)* catalyst fumes; *(D)* tongue depressor; *(E)* screen with paper towel; *(F)* catalyst.

Fig. 3.23. The computer-controlled vented warming oven used for Sabots A and B.

solution would deter the development of cracks in severely cracked wood, sabot A was placed into a container with a similar volume of CR-20 mixed with CT-32 catalyst, in a 1000:1 ratio. Like sabot B, sabot A was placed into a vented warming oven set at 71°C, and treated for 4.5 hours (fig. 3.23).

After treatment in their respective solutions, sabots A and B were removed from the oven and, still in their solutions, placed into a ventilated fume hood. Both artifacts were allowed to slowly return to room temperature over a period of approximately 3 hours. The sabots were then removed from their solutions and surface-wiped with soft, lint-free cloths. Inspection of the artifacts indicated that, just as before re-treatment, the surfaces of both sabots were mushy in texture; the cell structure of the wood was also badly deteriorated. Each sabot was then placed on a tray in the fume hood and allowed to air-dry overnight in flowing, fresh air. Changes in coloration were noticed immediately. During the first 30 minutes of air-drying, the sabots became lighter in color and more natural in appearance.

Observations

After 18 hours of exposure to fresh air, both sabots were measured, weighed, and photographed. The most notable change was in coloration. In both cases, the overall color of the artifacts had turned from a dark brown-black to a lighter, sandy brown. Neither sabot was greasy to the touch, and in both cases, the surfaces were dry and natural in appearance. Prior to re-treatment, the surfaces of both artifacts looked as though they had experienced some dehydration, either in the field during excavation or through too rapid treatment in the lab using PEG. The re-treatment process to remove free-flowing PEG from the artifacts did not appear to have exacerbated the problem.

Small nails had been used to fasten the iron bands that held the spherical shot in the sabot when the ammunition was manufactured. In both artifacts, small cracks centered around these nail holes were noted prior to re-treatment. In all cases, the cracks in sabot A were more visible after re- treatment, with one crack appearing to be slightly larger. The cracks in sabot B, however, reacted differently to treatment. In most cases, the cracks remained small and difficult to see. One small crack, which ran from a nail hole upward to the thin edge of the sabot, was not visible after re-treatment, suggesting that the wood may have swelled slightly, closing the crack. Pre- and post-re-treatment data for both artifacts are shown in table 3.3.

Conclusions

Clearly, less shrinkage occurred in sabot B than in sabot A. Minimal changes in cracks associated with nail holes running through the sides of sabot B suggest that the presence of catalyst mixed with CR-20 assisted in stabilizing the wood. The presence of catalyst also appears to have prevented the slightly higher rates of shrinkage noted in sabot A.

Table 3.3 Pre– and Post–re-treatment Data for Sabots A and B

Sample	*Pretreatment*	*Posttreatment*	*% Change*
Sabot A			
Top diameter (in cm)	11.050 (1)	11.01	–0.36
Top diameter (in cm)	11.026 (2)	11.00	–0.20
Base diameter (in cm)	9.72	9.70	–0.21
Height (in cm)	4.88	4.67	–4.31
Weight (in g)	266.20	238.40	–9.58
Sabot B			
Top diameter (in cm)	10.986 (1)	11.04	0.49
Top diameter (in cm)	10.984 (2)	11.01	0.24
Base diameter (in cm)	10.00	10.00	0.00
Height (in cm)	4.65	4.65	0.00
Weight (in g)	262.20	245.00	–6.56

On the basis of this experiment, however, it is not evident what role the presence of CR-20 played in accounting for the higher percentage of weight reduction for sabot A.

Bilz et al. have noted that PEG degrades even at temperatures as low as 39°C.[7] Numerous museums have approached the Archaeological Preservation Research Laboratory asking for directed research to prevent the obvious deterioration of PEG-treated artifacts that have been curated, even in the best of controlled environments at room temperature. Data from recent experimentation suggest that PEG extraction and polymerization can be successfully conducted at room temperature. The process takes longer, but the room temperature treatments help to minimize cracking and surface-checking. Additional experiments should be conducted to take into account the notable factors and results included in the Bilz et al. study.

As Bilz et al. and others have noted, PEG does degrade by random chain scission, especially in an anaerobic environment. Experiments conducted on similarly unstable polymers (silicone oils) have indicated that the use of antioxidants such as hydroxy anisole may inhibit aspects of oxidation and silicone oil degradation. Substantial research needs to be conducted on the use of antioxidants to help stabilize polymer-bulking media.

Initial tongue depressor experiments indicate that Passivation chemistries allow the conservator to remove free-flowing PEG from wood, and where applicable, effectively reduce treatment-associated swelling that can occur in PEG-treated artifacts. Passivation chemistries appear to hold some potential for removing free-flowing PEG from organic materials and for stabilizing artifacts that have been treated using PEG. Research at the Archaeological Preservation Research Laboratory continues to define experiments that may help us understand new methods of working with PEG and other polymer processes.

Case Study: Treatment of Waterlogged Wood Using Hydrolyzable, Multifunctional Alkoxysilane Polymers

Three groups of waterlogged wood samples were used in this experiment. Group 1 samples consisted of waterlogged tongue depressors. Group 2 samples were sectioned from a large piece of archaeological wood taken from marine excavations of the 1692 community of Port Royal, Jamaica. Group 3 samples consisted of 18 treenails (wooden dowels) extracted from the frames and large timbers of the seventeenth-century shipwreck *La Belle.* Each of these hand-carved pieces of wood was similar in diameter (mean 26.66 mm) and length (mean 128.70 mm). Many of their surfaces bore diagnostic tool marks.

Nuclear magnetic resonance (NMR) spectra of waterlogged wood treated with alkoxysilane polymers indicates that, in an aqueous environment, CR-20 hydrolyzes to form a triol. The triol self-condenses to form a range of polymers in the ^{29}Si spectra. The primary silicon has a methyl group and three siloxy bonds (i.e., So-O-Si). The second silicon environment has only two bonds. The third is formed to the methyl group, while the fourth bond is to the hydroxy group as in the ^{1}H spectrum (fig. 3.24). One goal of this experiment was to determine whether these polymers were sufficient to maintain the physical attributes, cell structure, and aesthetics of group 1, 2, and 3 wood samples.

NMR analysis was also conducted to determine whether the waterlogged tongue depressors were sufficiently degraded to provide a valid substitute for archaeological wood in such an experiment. Spectra of the group 1, 2, and 3 wood samples were nearly identical to spectra reported by Michael Wilson et al. in *The Degradation of Wood in Old Indian Ocean Shipwrecks.*[8] To determine the physical integrity of the group 1 wood samples, control samples were oven-dried over a 24-hour period. In all cases, the degree of warping and shrinkage indicated that they responded similarly to waterlogged archaeological wood.

Classroom conservation experimentation demonstrated that waterlogged tongue depressors proved to be good indicators of the effectiveness of traditional treatment methods, such as polyethylene glycol, acetone rosin, and sucrose. This same type of waterlogged wood appeared to work equally well as indicators of the effectiveness of polymer preservation treatment strategies.

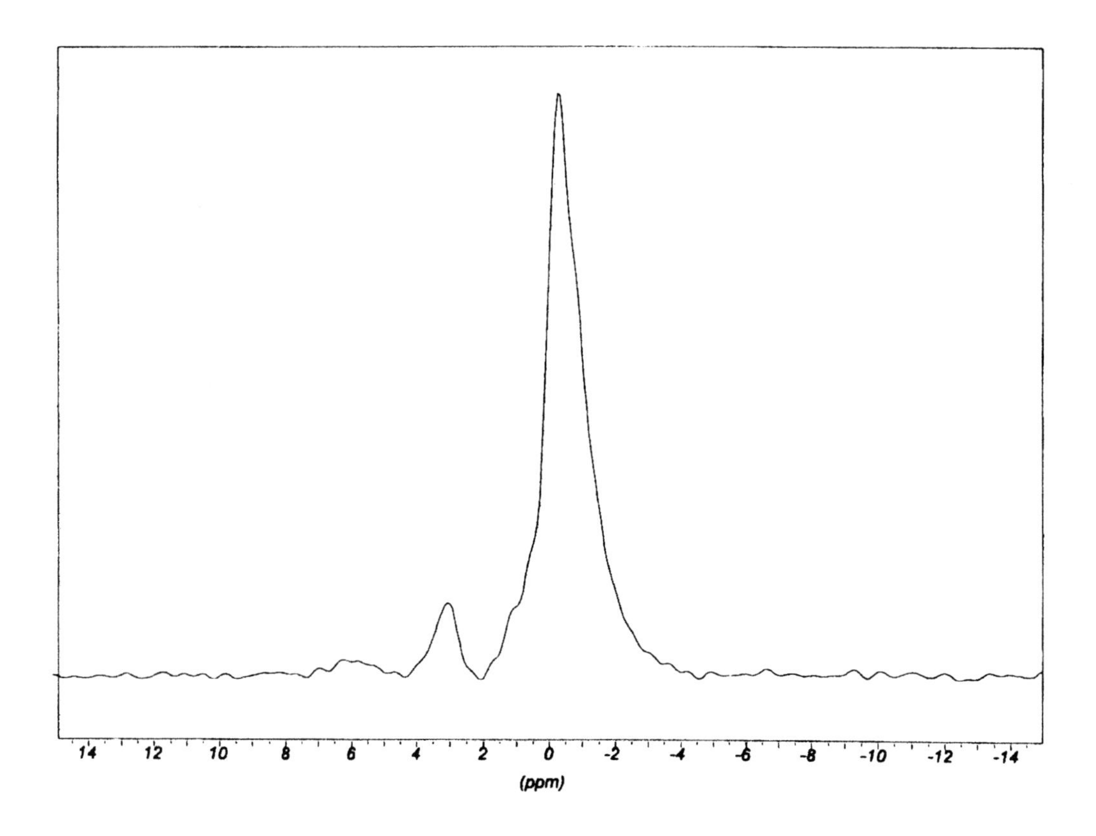

Fig. 3.24. ^{1}H CRAMPS spectrum of the solid polymer formed from the CR-20.

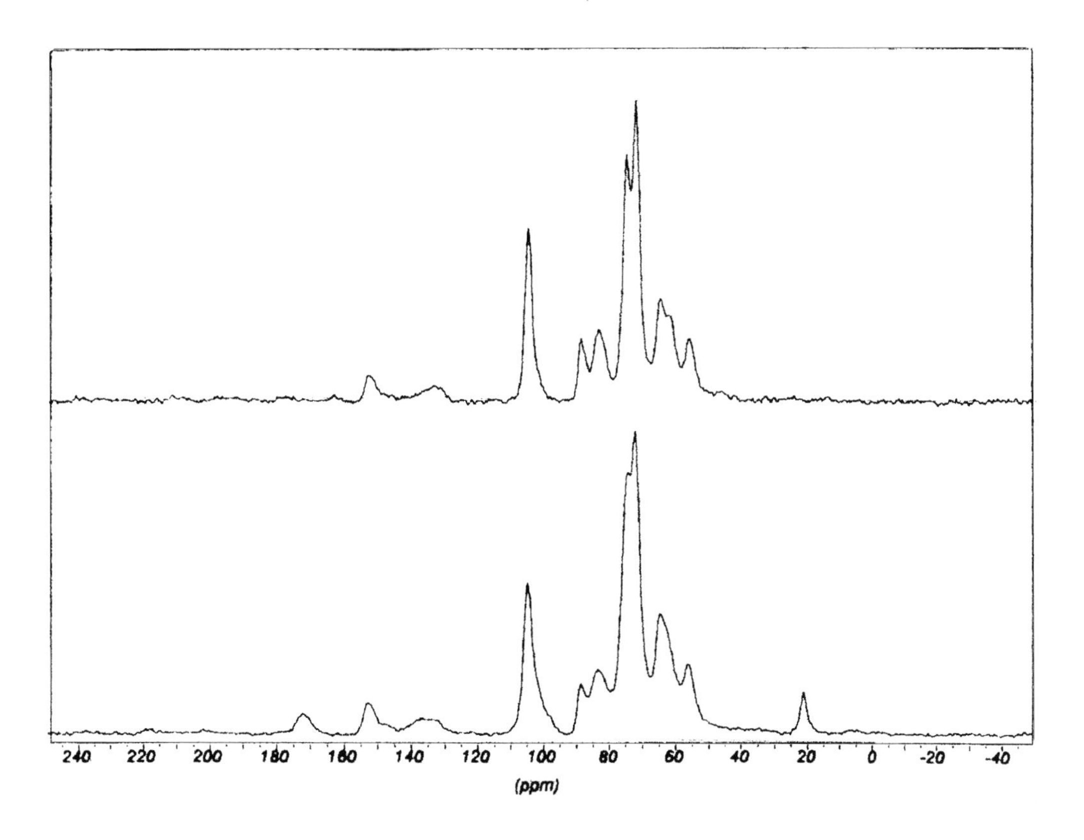

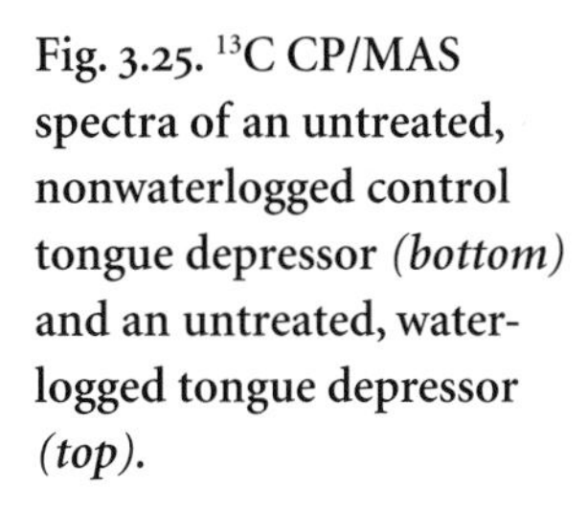

Fig. 3.25. ^{13}C **CP/MAS spectra of an untreated, nonwaterlogged control tongue depressor *(bottom)* and an untreated, waterlogged tongue depressor *(top)*.**

Materials—Group 1 Waterlogged Tongue Depressors

Waterlogged tongue depressors were chosen for group 1 samples because they are easily obtained and relatively uniform in dimension, grain, and color. To create a supply of waterlogged wood, thousands of white birch *(Betula papyrifera)* tongue depressors were placed into one-gallon glass jars filled with tap water. The jars were then sealed and stored in a cabinet.

For this experiment, hundreds of tongue depressors, which had been immersed in tap water for a period of nine years, were emptied into a plastic vat and rinsed in running tap water for 2 hours. Ten tongue depressors were selected randomly from the rinse vat, designated as air-dry samples, and used to determine the mean water content of the group 1 samples. Eighteen additional tongue depressors were randomly selected for treatment using acetone dehydration followed by acetone/CR-20 displacement.

The ^{13}C CP/MAS spectrum for one group 1 sample is illustrated at the top of figure 3.25. Long-term saturation in tap water altered the chemical structure of the wood, as evidenced by the loss of the acetate resonances at 22 and 174 ppm in comparison to the spectrum of a control tongue depressor that had not been waterlogged (fig. 3.25, *bottom*). The changes in these ^{13}C spectra of the birch wood tongue depressors are quite similar to those reported by Wilson et al. for oak wood from shipwrecks.[9] The ^{13}C spectral signature and the macroscopic observations of extensive warping and shrinking following air-drying suggest that these samples provide a suitable model for the analysis of waterlogged wood.

Materials—Group 2 Waterlogged Archaeological Wood

A small plank of wood recovered during archaeological excavations at the submerged site of seventeenth-century Port Royal, Jamaica, was selected for group 2 wood samples. The plank, measuring 12.32 cm wide, 14.73 cm long, and 1.87 cm thick, was cut into

Fig. 3.26. Surface characteristics, sections, and dimensions of the waterlogged plank used for the group 2 waterlogged wood samples. Top surface of the plank *(left)*; obverse surface *(right)*.

four sections. Due to its waterlogged state, the plank was very fragile. Sectioning was accomplished using a long scalpel blade. Figure 3.26 illustrates the dimensions of the group 2 wood samples.

Materials—Group 3 Waterlogged Treenails

Group 3 samples were treenails extracted from the timbers of a seventeenth-century shipwreck, *La Belle*. Each piece of wood was roughly carved and slightly tapered in shape. After desalination in freshwater baths for 24 months, 18 treenails were surface-dried with paper towels and then weighed, measured, and photographed (figs. 3.27 and 3.28). Sixteen of the treenails were then placed into a series of four ethanol baths, followed by a series of four acetone baths, each lasting two weeks. Eight of the treenails were picked randomly and immersed in CR-20. Eight were immersed in Q9-1315. The remaining two treenails were air-dried for 48 hours in a vented warming oven to determine percentage of water content.

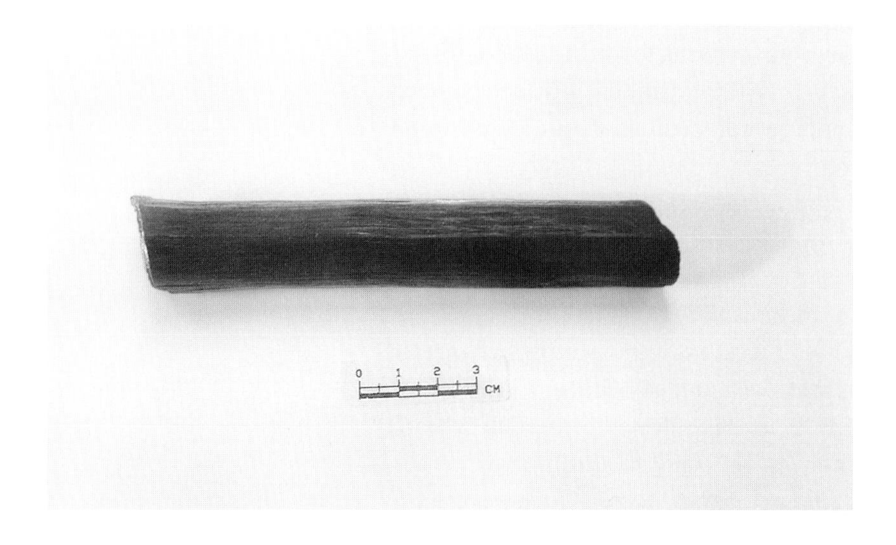

Fig. 3.27. Waterlogged treenail before air-drying.

Materials—Chemicals

CR-20 is a chemical monomer that reacts with water to form silanetriol and methanol. The silane in turn condenses with available hydroxyl groups or other silanol monomers to form siloxane resins. CR-20 consists of 97% methyltrimethoxysilane, 2% methyl alcohol, and 1% dimethyldimethoxysilane. The condensation product of CR-20 is a resin with a molecular weight of 226. Like CR-20, Q9-1315 is generally a clear liquid. However, because of the lower percentage of CR-20 and higher percentage of alcohols, evaporation during treatment is greater, and there are fewer hydroxyl groups and other silanol monomers available to form resins. The Q9-1315 solution is complex, and consists of 44% CR-20 (by weight), mixed with 50% methyl alcohol, 4% isopropyl alcohol, 1% ethyl alcohol, and 1% dimethyldimethoxysilane. Industrial-grade acetone, certified to be 99.78% free of water, was used for all dehydration processing.

Water Content in Group 1, 2, and 3 Wood Samples

Percentage of water content was calculated for each group of samples using the following equation:

$$\%H_2O = \frac{\text{weight of wet wood} - \text{weight of oven-dried wood}}{\text{weight of oven-dried wood}} \times 100$$

GROUP 1 WOOD

Ten group 1 waterlogged tongue depressors, designated as air-dry samples, were placed into a ventilated warming oven, set at 40°C, for 24 hours. The mean water content of the group 1 samples air-dried in the oven was 215.96%.

GROUP 2 WOOD

After 24 hours of air-drying in a vented warming oven, set at 40°C, W1 wood section weighed 4.4 g, a weight loss of 88.75%. Water content was calculated to be 788.64%. Because of the uniform thickness and condition of the wood, the water content calculation for W1 was assumed to reflect the general state of degradation of the other wood sections.

Prior to air-drying, W4, the second air-dried archaeological wood sample, had a mass of 104.3 g. W4 was placed into a fume hood and allowed to air-dry for 36 hours at ambient pressure and a constant room temperature of 76°F (24.4°C). After drying, it weighed 10.3 g, a weight loss of 90.12%, with a calculated water content of 912.62%.

Fig. 3.28. Treenail following air-drying. Note that this treenail splintered into three sections.

GROUP 3 WOOD

Two waterlogged treenails, weighing 82.1 g and 68.2 g wet weight, were allowed to air-dry in a well-ventilated fume hood for 48 hours. After air-drying, the treenails weighed 17.85 g and 17.80 g, a weight loss of 78.26% (359.94% water content) and 73.90% (283.15% water content), respectively.

Treatment Methods

GROUP 1 WOOD

Eighteen group 1 waterlogged tongue depressors were surface-dried with paper towels and labeled incrementally with a felt tip pen. Each sample's weight, length, width, and thickness were recorded.

The samples were placed into a large beaker containing one liter of fresh, industrial-grade acetone. After 24 hours, they were transferred to a second beaker, containing one liter of fresh acetone, and dehydrated for an additional 24 hours. After 48 hours of dehydration, the group 1 wood was transferred into a beaker containing one liter of CR-20. The beaker was placed into a desiccator vacuum chamber, where the wood was held in a reduced pressure environment of 5333.33 Pa (40 Torr) for 6 hours. The wood was then allowed to sit in the solution at ambient pressure and room temperature for 18 hours before being removed from the CR-20 solution and placed on paper towels in a vented fume hood to air-dry for 2 hours.

The samples were then placed in close proximity to an aluminum weighing dish containing 15 g of tap water and enclosed in a large Ziploc bag, thus creating a closed, humid environment. After 24 hours of exposure to water vapor, the wood was removed from the bag and allowed to air-dry in a fume hood before measurements and assessment of the treatment process were made.

GROUP 2 WOOD

Two sections of the plank, W2 and W3, were chosen for treatment with alkoxysilanes. W2 was placed into a beaker containing one liter of industrial-grade acetone and dehydrated at ambient pressure and room temperature for 24 hours. The wood was then placed in fresh acetone for an additional 24 hours of dehydration. After a total of 48 hours of dehydration, W2 was transferred to a beaker containing one liter of CR-20 and placed into a desiccator vacuum chamber. A reduced pressure of 5333.33 Pa (40 Torr) was applied for 10 hours. The wood was left in the solution at ambient pressure and room temperature for an additional 12 hours. After acetone/CR-20 displacement, W2 was removed from the CR-20 and placed into a Ziploc bag. An aluminum weighing dish, containing 20 g of tap water, was placed inside the bag, in close proximity to the CR-20-treated wood. The bag was sealed and the wood was allowed to sit for 18 hours. The wood was then removed from the bag and placed in a ventilated fume hood for an additional 24 hours of exposure to fresh air. After air-drying, the wood was weighed and measured.

Like W2, W3 was placed into a beaker containing one liter of industrial-grade acetone and dehydrated at ambient pressure and room temperature for a total of 48 hours. The wood was then transferred to a beaker containing one liter of Q9-1315 and placed into a desiccator vacuum chamber. A reduced pressure of 5333.33 Pa (40 Torr) was applied for 10 hours. The wood was then left in solution at ambient pressure and room temperature for an additional 12 hours. After acetone/Q9-1315 displacement, W3 was removed from the Q9-1315 solution and placed in a Ziploc bag with an aluminum weighing dish containing 20 g of tap water. The bag was sealed and the wood was allowed to sit for 18 hours. The wood was then removed from the bag and placed in a ventilated fume hood for an additional 24 hours of exposure to fresh air. After air-drying, the wood was weighed and measured.

GROUP 3 WOOD

Sixteen treenails were first dehydrated in a series of three ethanol baths, each lasting one week. Dehydration was then continued in a series of three acetone baths, changed at two-week intervals. For the last 10 hours of dehydration, the samples were placed into a large vacuum chamber and treated at a reduced pressure of 5333.33 Pa (40 Torr). During this phase of dehydration, the samples were monitored closely to ensure that they remained immersed in acetone.

Eight of the treenails were then randomly selected and carefully transferred to a large beaker containing one liter of CR-20. The remaining treenails were transferred into a beaker containing one liter of Q9-1315. Both beakers were placed into a vacuum chamber and treated at a reduced pressure of 5333.33 Pa (40 Torr) for 24 hours. The treenails were then stored at ambient pressure in their respective polymer solutions for an additional seven days. All the treenails were removed from their solutions, surface-dried with paper towels, and then placed into a fume hood, where they were allowed to air-dry for 24 hours.

Observations: Group 1 Wood

Prior to any treatment, the mean wet weight of the group 1 CR-20-treated wood samples was 5.63 g. After treatment, the mean weight was reduced to 2.59 g, a mean reduction of 53.83% (table 3.4).

The mean width of the 10 waterlogged tongue depressor samples that were simply air-dried was reduced from 17.89 mm to 13.67 mm, a 23.58% reduction (table 3.5). The mean length of these samples was reduced from 152.37 mm to 151.74 mm, a 0.71% reduction (table 3.6). The mean thickness of these samples decreased from 0.17 cm to 0.15 cm, a reduction of 11.76%. One control tongue depressor and several air-dried group 1 pieces of wood are illustrated in figure 3.29. Shrinkage, distortion, and color change are evident in these examples. After air-drying, the color of the waterlogged tongue depressors had shifted from their natural light yellow brown (10 YR/8/2 Munsell) to a light gray brown color (2.5Y/7/2 Munsell) or a darker brownish gray (10 YR/6/2 Munsell).

In contrast, the dimensions and aesthetic attributes of all the group 1 CR-20-treated wood samples were well maintained after treatment. Changes in length were minimal, decreasing only 0.41% after CR-20 treatment (table 3.7). Changes in width were noticeably higher, with an average posttreatment reduction of 11.11% (table 3.8). One control tongue depressor and nine CR-20-treated wood samples are illustrated in figure 3.30.

Cross section samples of control, untreated waterlogged wood, air-dried wood, and CR-20- treated tongue depressors were analyzed using an environmental scanning electron microscope (ESEM). For analytical consistency, photographs of each sample were recorded at 1000-X magnification. Cell shape retention, cell wall integrity, and general appearance were used to assess the efficacy of the treatments.

In figure 3.31, the left image is a 1000-X magnification of the cross-sectional surface of an untreated birch *(Betula papyrifera)* control tongue depressor showing uniformly shaped, thick-walled tracheids. In contrast, the tracheids in the waterlogged wood sample *(right)* are irregular in shape, with deterioration of the middle lamella. Figure 3.32 shows two views of the microstructure of a group 1 CR-20-treated sample of wood. Cell wall collapse is negligible, and there is very little distortion and structural loss of middle lamella. In figure 3.33, two cross-sectional views indicate that, after air-drying, the cell structure of the group 1 waterlogged wood samples collapsed, causing extreme shrinkage and distortion of the wood.

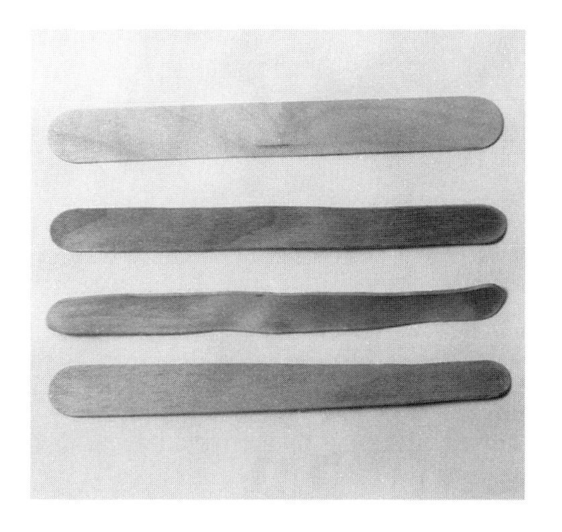

Fig. 3.29. Control tongue depressor *(top)*; three air-dried waterlogged tongue depressors *(below)*.

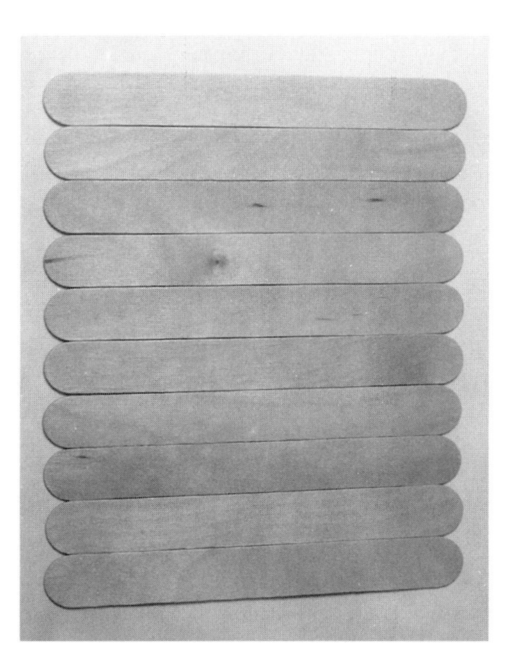

Fig. 3.30. Control tongue depressor *(top)*; nine CR-20-treated, waterlogged tongue depressors *(below)*.

Table 3.4 Weight Changes for Group 1 CR-20-Treated Tongue Depressor Samples

TD	*Wet Weight (in g)*	*Posttreatment Weight (in g)*	*% Change*
1	5.0	2.7	–46.00
2	5.8	2.6	–55.17
3	5.3	2.4	–54.72
4	5.5	2.0	–63.64
5	5.7	2.3	–59.65
6	5.9	3.0	–49.15
7	5.4	2.5	–53.70
8	5.8	2.6	–55.17
9	5.7	3.0	–47.37
10	5.7	2.8	–50.88
11	5.5	2.5	–54.55
12	5.5	2.5	–54.55
13	5.8	2.7	–53.45
14	5.7	2.6	–54.39
15	5.8	2.6	–55.17
16	5.7	2.6	–54.39
17	5.6	2.5	–55.36
18	5.8	2.8	–51.72
	Mean Wet Weight	*Mean Post-treatment Weight*	*Mean % Change*
	5.63	2.59	–53.83

Table 3.5 Width Change of Air-Dried Waterlogged Tongue Depressors

TD	*Wet Width (in mm)*	*Air-Dry Width (in mm)*	*% Change*
1	17.90	13.47	–24.75
2	17.87	14.32	–19.87
3	17.87	13.24	–25.91
4	17.91	13.43	–25.01
5	17.90	14.60	–18.44
6	17.88	12.90	–27.85
7	17.87	14.36	–19.64
8	17.89	13.53	–24.37
9	17.90	14.25	–20.39
10	17.91	12.61	–29.59
	Mean Wet Width	*Mean Air-Dry Width*	*Mean % Change*
	17.89	13.67	–23.58

Table 3.6 Change in Length of Air-Dried Waterlogged Tongue Depressors

TD	*Wet Length (in mm)*	*Air-Dry Length (in mm)*	*% Change*
1	152.50	151.08	–0.93
2	152.22	150.88	–0.88
3	152.47	150.86	–1.06
4	152.26	151.03	–0.81
5	152.48	150.97	–0.99
6	152.51	151.09	–0.93
7	152.59	151.77	–0.54
8	152.48	151.11	–0.90
9	152.35	151.76	–0.39
10	152.34	151.62	–0.47
	Mean Wet Length	*Mean Air-Dry Length*	*Mean % Change*
	152.37	151.74	–0.71

Table 3.7 Changes in Length for Group 1 CR-20-Treated Tongue Depressor Samples

TD	*Wet Length (in mm)*	*Preserved Length (in mm)*	*% Change*
1	152.49	151.75	–0.49
2	152.46	151.60	–0.56
3	152.34	152.04	–0.20
4	151.97	151.17	–0.53
5	152.48	152.03	–0.30
6	152.58	152.10	–0.31
7	151.82	151.33	–0.32
8	152.48	151.94	–0.35
9	152.68	151.96	–0.47
10	152.50	151.97	–0.35
11	152.59	151.55	–0.68
12	152.50	151.65	–0.56
13	152.46	151.97	–0.32
14	152.21	151.58	–0.41
15	152.58	151.99	–0.39
16	152.56	151.85	–0.47
17	152.51	152.14	–0.24
18	151.48	150.78	–0.46
	Mean Wet Length	*Mean Post-treatment Length*	*Mean % Change*
	152.37	151.74	–0.41

Table 3.8 Changes in Width for Group 1 CR-20-Treated Tongue Depressor Samples

TD	*Wet Width (in mm)*	*Post Treatment Width (in mm)*	*% Change*
1	19.71	17.59	−10.76
2	19.78	17.95	−09.25
3	19.66	17.35	−11.75
4	19.70	17.95	−08.88
5	19.80	17.70	−10.61
6	19.00	17.79	−11.05
7	19.72	17.93	−09.08
8	19.80	17.33	−12.48
9	19.77	17.95	−09.21
10	19.75	17.94	−09.17
11	19.79	17.62	−12.32
12	19.64	17.24	−12.22
13	19.70	17.05	−13.45
14	19.87	17.09	−13.99
15	19.59	17.79	−10.12
16	19.71	17.61	−10.65
17	19.70	17.18	−12.79
18	19.90	17.48	−12.16
	Mean Wet Width	*Mean Post-treatment Width*	*Mean % Change*
	19.76	17.86	−11.11

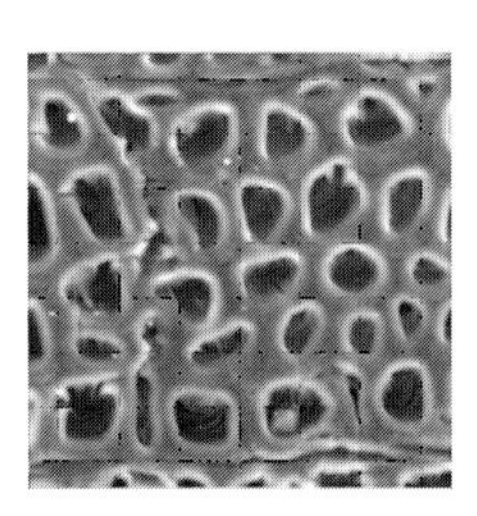
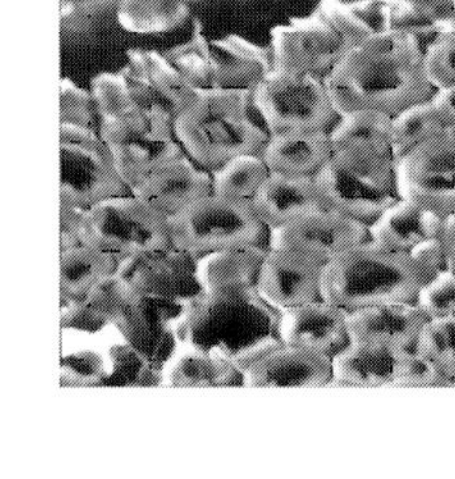

Fig. 3.31. Control group 1 tongue depressor *(left)* and an image of a waterlogged group 1 tongue depressor *(right)*, both viewed at 1000-X magnification.

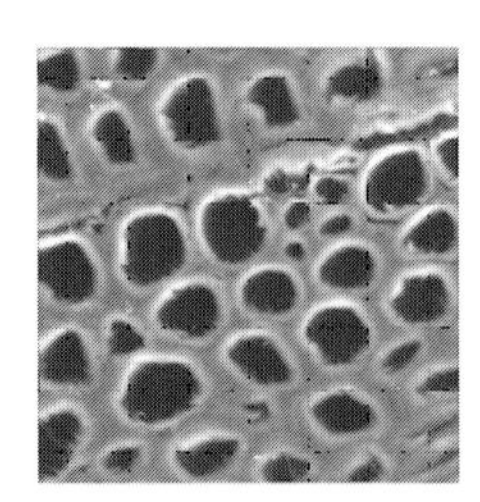
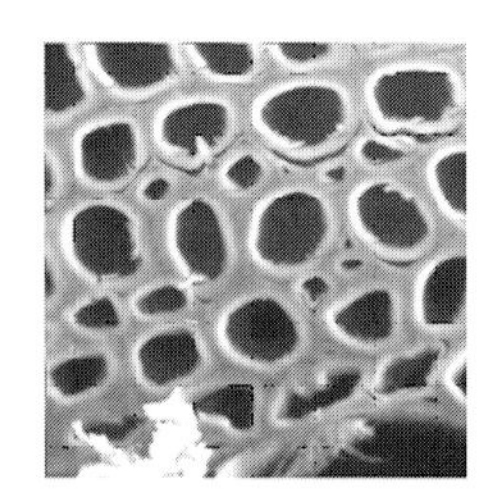

Fig. 3.32. Two cross-sectional microscopic views (1000-X magnification) of a group 1 tongue depressor treated with CR-20.

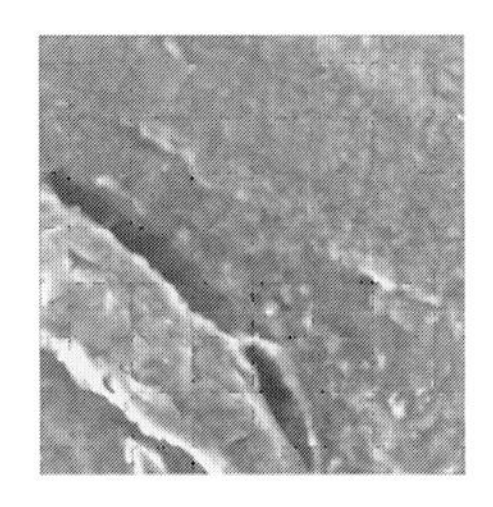
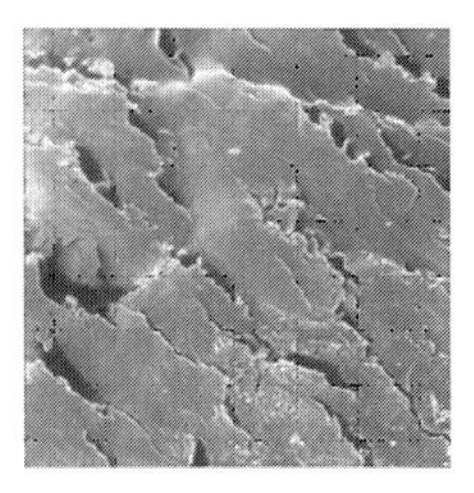

Fig. 3.33. Two cross-sectional views of group 1, air-dried wood samples. In both images, cellular distortion and collapse are apparent, resulting in extreme warping and distortion.

Observations: Group 2 Wood

Prior to air-drying in a warming oven, W1 weighed 39.1 g and measured 2.00 cm wide, 1.87 cm thick, and 12.20 cm long. Because W1 was cut thinner than W2, W3, and W4, it was very flexible and delicate. After 24 hours of air-drying, W1 was reduced from its wet weight of 39.1 g to 4.4 g, a decrease of 88.75%. The water content of this sample was 788.64%. When removed from the warming oven, the wood had completely collapsed and fragmented into five significant sections, which precluded length and width measurements. Prior to treatment, W1 was 1.87 cm thick. Mean thickness of W1 wood fragments after treatment was 0.42 cm, a reduction of 77.54%.

W2 was sectioned from the plank adjacent to W1. This section of wood had a wet weight of 115.7 g and measured 4.13 cm wide, 1.86 cm thick, and 14.68 cm long. W2 was designated for treatment in CR-20, after dehydration in acetone. After treatment in CR-20, W2 appeared uniformly dry and light in color. Only slight dimensional changes were noted. The wet weight of W2 was reduced from 115.7 g to 28.7 g, a decrease of 75.20%. W2 measured 4.13 cm wide prior to treatment. After treatment, it measured 3.96 cm, a reduction of 4.12%. Thickness of W2 prior to treatment was 1.86 cm. After treatment, it measured 1.85 cm thick, indicating that little or no significant change had occurred in thickness. Before treatment, W2 measured 14.68 cm long. After treatment, it measured 14.63 cm long, a loss of 0.34%.

W3 had a wet weight of 91.3 g. Its wet measurements were 3.14 cm wide, 1.84 cm thick, and 14.73 cm long. W3 was designated for treatment in Q9-1315, after an initial dehydration in acetone. After treatment in Q9-1315, the weight of W3 decreased from 91.3 g. to 26.0 g, a loss of 71.52%. The width of the sample decreased from 3.14 cm to 2.98 cm, a reduction of 5.10%. The thickness of W3 was reduced from 1.84 to 1.62 cm, a decrease of 11.96%. The wet length of W3 was 14.73 cm. After treatment, the wood measured 14.54 cm, a reduction in length of 1.29%.

The remaining section of wood, W4, was designated for air-drying. Prior to treatment, its wet weight was 104.3 g. Its wet measurements were 3.05 cm wide, 1.87 cm thick, and 12.27 cm long. W4 was the largest of two air-dried samples from the original plank of waterlogged wood. Unlike W1, W4 was similar in size to the sections of wood treated with alkoxysilane polymers. Prior to treatment, its wet weight was 104.3 g. After treatment, it weighed 10.3 g, a loss of 90.13%. Its width decreased from 3.05 cm to 2.55 cm, a reduction of 16.39%. The thickness of the sample fell from 1.87 cm to approximately 0.72 cm, a reduction of 61.50%. The posttreatment length of W4 was difficult to determine. During treatment, the sample splintered into six large sections, each suffering gross distortion of its edges. Prior to air-drying, the sample measured 12.27 cm at its longest point. After treatment, W4 measured approximately 10.41 cm long, a loss of 15.16%. The pre- and posttreatment measurements of all four group 2 samples are summarized in table 3.9.

After 24 hours of treatment, the W1 sample of wood had deteriorated into a pile of splinters, precluding a comparison of pre- and posttreatment dimensions and aesthetic qualities. The wood section was computed to have a water content of 788.64%.

Prior to treatment, the W2 wood was dark brown (10 YR/3/3 Munsell). After treatment, it was a light gray brown (10 YR/6/2 Munsell). The surfaces showed no signs of checking and the wood looked very natural. After treatment, however, W2 was very light in weight, but the preserved wood withstood extensive handling with no signs of deterioration or wear.

After treatment, W3 was also aesthetically pleasing. Prior to treatment, W3 was dark brown (10 YR/3/3 Munsell). After treatment, the wood was slightly darker (10 YR/4/2 Munsell) than W2. Both W2 and W3 wood samples were natural in appearance, and the surfaces of both pieces of wood felt dry. Shrinkage of W3, however, was more of a problem.

Unlike W1, which was oven-dried for 24 hours, W4 was allowed to air-dry in a vented fume hood. W4 splintered into six large sections. Fragmentation of the wood was less severe than that of W1. Each of the W4 splinters of wood was distorted, making posttreatment measurement of dimensions difficult. Because shrinkage and fragmentation were extensive, it was impossible to fit the distorted sections of wood into a form that could be accurately measured. The posttreatment condition of the four wood samples is illustrated in figure 3.34.

Observations: Group 3 Wood

Prior to treatment, the mean percentage water content in the group 3 waterlogged treenails was 321.55%, appreciably lower than the water content of 788.64% calculated for the group 2 wooden plank samples. During

Table 3.9 Pre- and Posttreatment Measurement for Group 2 Wood Samples

Sample	*Pretreatment*	*Posttreatment*	*% Change*
W1[1]			
Weight (in g)	39.1	4.4	−88.75
Length (in cm)	12.20	—	—
Width (in cm)	2.00	—	—
Thickness (in cm)	1.87	0.42	−77.54
W2			
Weight (in g)	115.7	28.7	−75.20
Length (in cm)	14.68	14.63	−0.34
Width (in cm)	4.13	3.96	−4.12
Thickness (in cm)	1.86	1.85	−0.54
W3			
Weight (in g)	91.3	26.0	−71.52
Length (in cm)	14.73	14.54	−1.29
Width (in cm)	3.14	2.98	−5.10
Thickness (in cm)	1.84	1.62	−11.96
W4[2]			
Weight (in g)	104.3	10.3	−90.13
Length (in cm)	12.27	10.41	−15.16
Width (in cm)	3.05	2.55	−16.39
Thickness (in cm)	1.87	0.72	−61.50

[1] Fragmentation precluded posttreatment length and width measurements.
[2] Fragmentation precluded exact posttreatment length measurement; the value is an approximation.

initial cleaning and desalination, the treenails were found to be less spongy than the sections of plank, and the wood was noticeably harder. Many of the ends of the treenails had either been broken or splayed under the force of being removed from the ship's timbers.

Fig. 3.34. Group 2 wood after treatment. *W1*, oven-dried wood; *W2*, CR-20-treated wood; *W3*, Q9-1315-treated wood; *W4*, air-dried wood in vented fume hood. Note the comparatively lighter color of *W2*.

Group 3 wood samples treated with CR-20 alkoxysilane polymers experienced only slight changes in posttreatment weight, length, and diameter compared to the group 3 Q9-1315-treated wood. After treatment, the mean weight of CR-20-treated treenails was 40.48 g, a reduction of 45.74%. Treenails preserved using Q9-1315 polymer had a mean weight of 39.75 g after treatment, a reduction of 43.76% (table 3.10). Group 3 treenails preserved using CR-20 had a mean reduction in length of 0.36%. The change in length was greater for Q9-1315-treated treenails, with a mean posttreatment length of 121.76 mm. This represents a shrinkage of 0.49% shrinkage (table 3.11). Similarly, the mean change in diameters for the group 3, CR-20-treated nails was substantially less than the mean change in diameter for Q9-1315-treated treenails. The posttreatment mean diameter for CR-20-treated wood was 25.75 mm, a reduction of 3.42%. Q9-1315-treated treenails had a mean loss in posttreatment diameter of 9.36% (table 3.12).

The most noticeable difference between the CR-20- and Q9-1315-treated treenails was in coloration. In all cases, the CR-20-treated wood was much lighter in color. Q9-1315-treated wood tended to be darker, with fewer wood grains and surface features visible.

Discussion

NMR spectral analysis, ESEM analysis, and empirical data indicate that the structural integrity of the group 1 tongue depressors was sufficiently degraded that the wood samples can be used to evaluate preservation treatments for waterlogged timbers from shipwrecks. Group 1 wood samples also provide a reasonably homogenous source of wood that allows quantifiable and qualitative analysis of the efficacy of consolidants being tested for use in conserving waterlogged wood. Regularity of size and species, as well as the availability of nonwaterlogged control samples, make the group 1 wood samples invaluable for experimentation. Because of inherent inconsistencies of waterlogged archaeological wood, similar comparative data cannot be derived for archaeological samples.

Air-dried group 1 wood samples experienced a mean weight loss of 54.00%, a mean reduction in sample width of 23.58%, and a mean reduction in length of 0.71%. All the samples were warped and twisted after air-drying.

CR-20-treated group 1 wood samples were generally well preserved. Mean reduction in length was 0.41% after treatment. Mean reduction in width was 11.11%. These figures may seem high, but a comparison of the posttreatment dimensions of the CR-20-treated samples shows that they are nearly identical to the untreated, control depressors, indicating that the treated wood was restored to dimensions nearly identical to those of the

Table 3.10 Changes in Weight for Group 3 Treenail Artifacts

Treenail	*Treatment*	*Wet Weight (in g)*	*Posttreatment Weight (in g)*	*% Change*
1	Air-dry	68.2	17.80	−73.90
2	Air-dry	82.1	17.85	−78.26
	Mean (Air-dry)	75.15	17.83	−76.08
3	CR-20	83.8	44.13	−47.34
4	CR-20	59.8	32.42	−45.79
7	CR-20	66.2	37.99	−42.61
9	CR-20	65.7	35.36	−46.18
10	CR-20	91.1	50.75	−44.29
12	CR-20	67.4	38.32	−43.15
15	CR-20	75.5	38.67	−48.78
18	CR-20	84.2	44.00	−47.74
	Mean (CR-20)	74.21	40.48	−45.74
5	Q-9-1315	69.7	37.83	−45.72
6	Q-9-1315	62.2	32.86	−47.17
8	Q-9-1315	77.8	40.85	−47.49
11	Q-9-1315	64.4	36.59	−43.18
13	Q-9-1315	82.1	46.94	−42.83
14	Q-9-1315	55.5	33.43	−39.77
16	Q-9-1315	91.5	53.90	−41.09
17	Q-9-1315	63.5	36.32	−42.80
	Mean (Q- 9-1315)	70.84	39.75	−43.76

Table 3.11 Changes in Length for Group 3 Treenail Artifacts

Treenail	*Treatment*	*Pretreatment Length (in mm)*	*Posttreatment Length (in mm)*	*% Change*
1	Air-dry	106.5	92.53	−13.12
2	Air-dry	145.39	122.69	−15.61
	Mean (Air-dry)	125.95	107.61	−14.37
3	CR-20	140.84	140.30	−0.38
4	CR-20	115.43	115.06	−0.32
7	CR-20	148.55	148.18	−0.25
9	CR-20	111.36	110.74	−0.56
10	CR-20	122.47	122.05	−0.34
12	CR-20	108.61	108.11	−0.46
15	CR-20	144.59	144.29	−0.21
18	CR-20	137.75	137.27	−0.35
	Mean (CR-20)	128.70	128.24	−0.36
5	Q-9-1315	128.73	127.89	−0.65
6	Q-9-1315	113.28	112.76	−0.46
8	Q-9-1315	121.91	121.32	−0.48
11	Q-9-1315	115.53	114.86	−0.58
13	Q-9-1315	130.18	129.62	−0.43
14	Q-9-1315	124.13	123.62	−0.41
16	Q-9-1315	142.42	141.68	−0.52
17	Q-9-1315	102.68	102.29	−0.38
	Mean (Q-9-1315)	122.36	121.76	−0.49

Table 3.12 Changes in Diameter for Group 3 Treenail Artifacts

Treenail	*Treatment*	*Pretreatment Diameter (in mm)*	*Posttreatment Diameter (in mm)*	*% Change*
1	Air-dry	26.85	22.32	−16.87
2	Air-dry	24.87	20.93	−15.84
	Mean (Air-dry)	25.86	21.63	−16.36
3	CR-20	27.45	26.29	−4.23
4	CR-20	25.05	24.14	−3.63
7	CR-20	26.11	25.44	−2.57
9	CR-20	27.92	27.14	−2.79
10	CR-20	27.38	26.39	−3.62
12	CR-20	26.75	25.65	−4.12
15	CR-20	25.20	24.36	−3.33
18	CR-20	27.39	26.56	−3.03
	Mean (CR-20)	26.66	25.75	−3.42
5	Q-9-1315	27.01	23.81	−11.85
6	Q-9-1315	26.83	24.27	−9.54
8	Q-9-1315	29.30	26.60	−9.22
11	Q-9-1315	26.77	24.82	−7.28
13	Q-9-1315	27.02	24.15	−10.63
14	Q-9-1315	22.31	20.68	−7.31
16	Q-9-1315	27.55	25.26	−8.31
17	Q-9-1315	27.58	24.62	−10.74
	Mean (Q-9-1315)	26.80	24.28	−9.36

control wood samples. Swelling that occurred during the waterlogging process had been greatly reduced after the wood was treated in CR-20. ESEM evaluation of these samples confirms that cell dimensions and shapes were similar to those of the control wood samples. Slight shrinkage of the middle lamella was noted after treatment.

After treatment, all the group 1 tongue depressors were slightly gray brown in color (10YR 7/2 Munsell) as compared to the color of the control tongue depressors (10YR 8/4 Munsell). This color shift is the result of changes in the wood caused by long-term immersion in water.

Group 2 wood samples were highly degraded. The computed water content of the samples was between 788.64% and 912.62%, suggesting that water content was reasonably uniform throughout the plank. Oven-dried W1 and air-dried W4 were much darker in color than the polymer-treated sections, W2 and W3. This is the result of extreme cellular collapse and warpage that occurred as a result of air-drying. Both W1 and W4 disintegrated to the point that accurate physical measurements were impossible to obtain.

The conservation of W2 and W3 was considered successful as their physical dimensions, surface textures, and individual characteristics were accurately maintained. Lower rates of shrinkage were observed in the CR-20-treated W2 wood sample, and it was lighter in color than W3. The W3 sample was darker due to the slightly higher rate of shrinkage that occurred as a result of treatment in Q9-1315.

CR-20 is 97% pure, with a 3% addition of alcohols. Q9-1315 is a less-refined solution containing approximately 44% CR-20 mixed with organic solvents and trace amounts of dimethyldimethoxysilane. Accordingly, the resin-forming capabilities of the Q9-1315 solution are insufficient to preserve the dimensional characteristics of the wood. This is evident in the slightly higher rates of shrinkage for W3.

The diagnostic attributes of W2 and W3 were preserved because sufficient resins were formed as the result of condensation to prevent cellular collapse of the wood. The waterlogged tongue depressors and the archaeological wood samples preserved with the CR-20 solution, which has a higher percentage of hydrolyzable, multifunctional polymers, were the best-preserved specimens. In contrast, the Q9-1315 solution, containing a lower percentage solution of the same multifunctional alkoxysilane polymers, was insufficient to preserve the diagnostic attributes of the wood. The waterlogged tongue depressors, calculated to have a moisture content of 215.96%, were very well preserved with CR-20.

Wood treated using a higher percentage of CR-20 solution looks natural in color and texture following treatment. No surface-checking was noted in either the group 1 wood samples or the archaeological wood. During the waterlogging process, the natural color of the wood was altered, resulting in a slightly grayish cast. The wood, however, was dimensionally stable after treatment in the silane. Experimentation indicates that resins created as the result of condensation can preserve even badly waterlogged wood very well. As C. V. Horie and others have suggested, conservation strategies using silicone oils are not reversible.[10] However, over time, the solubility of many adhesives and consolidants currently in common use in conservation are also affected, rendering these techniques nonreversible as well.[11]

More experimentation is needed to determine if the addition of small percentages of silicone oils might increase the bulking ability of CR-20 and Q9-1315, effectively reducing shrinkage. Adding a small percentage of a low viscosity silicone oil to Q9-1315 may increase its bulking ability despite its high alcohol content, making it an effective treatment.

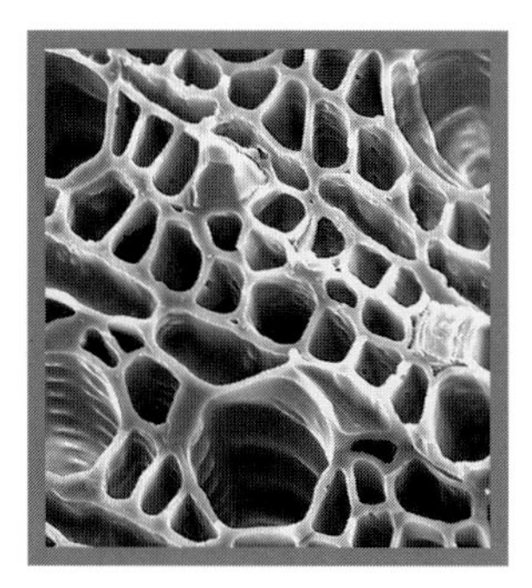

Leather Preservation

Leather recovered from archaeological sites presents a variety of challenges for the conservator. Regardless of their condition, leather artifacts reveal something about the technology and craftsmanship used in their creation. Tool marks may indicate the level of skill of the cobbler. Wear marks may help characterize the occupation and affluence of the owner.

Archaeological Leather

Whether from a terrestrial or marine site, archaeological leather is affected by numerous environmental factors that act to compromise its general stability. Leather recovered from land sites may be brittle if the relative humidity (RH) is less than 50%. Often, embrittled leather will crack and shrink excessively. In dry and desiccated leather, collagen fibers, which are polymers made up of amino acids, may become rigid due to a lack of moisture. These artifacts need careful rehydration to prevent continued cracking and deterioration. Dressings and topical applications of lanolin and cedarwood oils have been used to lubricate leather. To regain flexibility in dry leather, conservators have used glycerol, sorbitol, or a wide range of molecular weights of PEG.

Wet leather also requires careful treatment to prevent surface drying and hardening of collagen fibers. Excessive hydration may lead to the loss of tannins. In their absence, water lubricates collagen fibers and prevents shrinkage. Accordingly, PEG treatments based on the incremental addition of PEG to an aqueous solution in which the artifact is immersed have been the most common means of preserving waterlogged leather. In combination with freeze-drying technology, PEG treatments have been very successful in preserving the diagnostic attributes of waterlogged leather artifacts.

Mold, mildew, fungi, and exposure to ultraviolet light and airborne contaminants such as sulphur dioxide all act to undermine the integrity of leather. Singly, or in combination, these factors complicate the process of stabilization. The often delicate condition of archaeological leather necessitates that all destructive mechanisms be neutralized prior to treatment. In most cases, archaeological leather requires some degree of cleaning before conservation procedures are undertaken. Badly deteriorated leather, however, may be too friable to clean, and it may therefore be impossible to remove stains without causing extensive damage. In such cases, it may be best to focus on stabilizing the leather.

Variables to Consider

Many factors need to be considered before attempting to conserve leather. These factors may be beyond the abilities of the conservator to determine but nonetheless play a role in the state of deterioration affecting archaeological leather. Determining genus and species of the animal may be important when trying to assess the degree of flexibility and color desirable in a preserved piece of leather. The degree of use wear may also be an important consideration. Apart from being diagnostic features of the artifact, extensive scuff marks and surface abrasion may help determine a suitable treatment strategy.

Artifact association or proximity is another consideration when developing a conservation strategy for leather. Archaeological leather closely associated with large deposits of organic materials may be heavily degraded due to fungus and decay, whereas leather in close association with iron artifacts may be heavily concreted or impregnated with minerals and oxides.

Initial means of tanning used to preserve the leather may be a factor in an artifact's state of deterioration. This may also help determine viable treatment strategies for preserving the artifact. The earliest suspected means of preparing animal hides were concerned with slowing the processes of putrefaction. Besides hides, other organs including the stomach and bladder were utilized. Brain tanning was also common. In many cases, the brains of the animal were considered sufficient to tan the whole useable hide of the animal. Smoking and close skinning techniques, which removed thick layers of fat, were also commonly used to prepare hides. By removing the thick layers of subcutaneous tissue or using smoke to cure tissues, hides could be effectively preserved.

Historic period hides and leather were also commonly prepared using vegetable tanning methods. This incorporated the use of natural tannins found in the bark and roots of numerous plants. Hides were soaked in baths of bark and water in order to infuse them with tannins that combine easily with collagen and proteins. Tannins replace water in the hide and reinforce side chain structures, essentially cross-linking protein molecules, making them stable. They are effective antioxidants and help maintain flexibility of the treated hide. Vegetable tanning processes produce polyphenols, which readily form hydrogen bonds with carbon atoms associated with available hydroxyl groups within the leather. For this reason, vegetable tanned leather, in many cases, deteriorates rapidly as hydrolyzation interrupts the chemical equilibrium of the leather. Bark from oak trees is still commonly used for noncommercial tanning of hides; indeed, the bark and roots of many plants have tannins that are effective in stabilizing proteins.

For the most part, twentieth-century leather was tanned using chromium tanning processes; it is sometimes called mineral tanned leather. Chromium salts, which act to reinforce collagen molecular chains, also give leather a chemical resistance to many agents of decomposition including molds and fungi, as well as airborne contaminants producing sulfuric acid. In comparison, vegetable tanned hides are more susceptible to airborne contaminants since the cross-linking of proteins is not as complete as the chemical transformation and stabilization induced by chromium processing, which makes hides somewhat hydrophobic.

Regardless of the method used for tanning leather, hides share one common characteristic. Collagen fibers, which are fibrous proteins that give leather its flexibility, require natural lubrication for the hide to remain supple. These same fibers require chemical restructuring to prevent breakdown of the fibers and putrefaction. Hydrolyzation acts to break down collagen chains, thus reducing a skin's ability to remain supple once water has been removed from the hide. Waterlogged leather usually becomes increasingly soft as the process leaches tannins and fats from its matrix. Accordingly, determination of the physical state of deterioration can be difficult. High pH levels may indicate the presence of oxides in degraded leather. Numerous tests can be made on wet leather. Generally, these tests measure characteristics or qualitative aspects. Regardless of the state of the artifact, leather removed from a wet environment must remain wet until treated.

Cleaning

Waterlogged leather recovered from shipwreck sites can pose many challenges. Leather from these sites may be covered with concreted deposits, which are the product of oxidation of metals. Concreted material and impacted soil on leather from land sites should be removed carefully using soft wooden dowels and brushes. In some cases, leather may be too soft and fragile to clean. It is best to treat these artifacts with a minimum of cleaning and handling. Often, additional cleaning can be successfully accomplished after conservation, when the leather is in a more stable state.

To prevent structural damage to badly deteriorated leather, cleaning should be conducted using a soft brush, while the artifact is submerged in freshwater. Leather in good condition can be cleaned by placing it on a clean working surface and using a gentle, diverted flow of water to keep it wet. This allows the conservator to inspect the artifact more closely. A direct flow of water on leather should never be used since this may cause irreversible damage due to erosion and pitting of the artifact's surfaces.

While it is best to remove concretion from the surfaces of leather using soft wooden tools, heavier areas of concretion can be removed using an air scribe. With a gentle touch, using the vibration of the stylus to break apart hard deposits, this instrument can be more effective for removing thick, concreted materials than wooden dowels. Care must be taken, however, since these same vibrations can undermine the artifact's integrity. It is often better to hold a small piece of leather or place it on a soft bedding of cloth to absorb vibrations from the air scribe. This will substantially reduce stress on the artifact.

Many inexperienced conservators hold the tip of the air scribe at too sharp an angle, which directs the force of the stylus directly into the artifact. By working at a shallow angle to the line of the artifact, the force of the stylus is deflected, reducing stress. With practice, an air scribe or pneumatic chisel can be operated with finesse, allowing very fine and detailed mechanical removal of concreted materials from leather and other delicate artifacts.

Chemical Cleaning

Numerous organic acids can be used to remove corrosion and stains on leather recovered from marine environments. Citric acid, oxalic acid, disodium EDTA, tetrasodium EDTA, ammonium citrate, and dilute hydrochloric acid (1–4%) are a few of the most common treatments. Citric acid is a slow and gentle acid; hydrochloric acid tends to work more quickly. Whatever means are employed to remove corrosion, concretion, and stains from leather, it is essential to extensively rinse the artifact in fresh running water baths after treatment. If chemicals are not completely flushed from the matrix of the leather, there is a risk of chemical reactivity among tannins, fats, collagen fibers, and the acids and neutralizing agents used to treat the artifact. Accordingly, chemical cleaning of waterlogged or desiccated leather is often not the first choice of many conservators.

Treatment of Leather

It is difficult and risky to attempt conserving archaeological leather by controlled drying or freeze-drying methods without the aid of a replacement bulking agent to maintain the artifact's physical attributes. As water is removed from the artifact, collagen fibers are drawn together, causing hardening, shrinkage, and disfigurement. To prevent these problems, water within the matrix must be displaced with a suitable bulking agent that will enable the artifact to withstand the drying process. Over the last few decades, PEG has been widely used for the treatment of leather. Through incremental additions of PEG to water, the cells and voids in the matrix of leather can be sufficiently supported and lubricated to prevent the shrinkage and hardening that occur as the result of drying.

Anyone who has ever worked with wet leather has observed that the artifact remains soft and supple while immersed in water. When moved to a bath of organic solvent, such as ethanol, the leather becomes stiffer and harder to manipulate. This change in malleability occurs as water is replaced by the solvent. Progressive dehydration has been successfully used to conserve some wet leather artifacts. By taking the artifact through a series of baths, starting with ethanol and working up to fresh baths of acetone, it is possible to dehydrate leather into a relatively stable state.

The problem with such treatment strategies, as Cronyn notes, is that there is a chance of shrinkage and distortion in the artifact.[1] Degree of degradation, biological activity, rate of water-solvent exchange, variance in surface tension, and chemical reactivity with insoluble deposition on the surfaces of the artifact will affect dehydration and preparation of the leather for continuing conservation work.

PEG/Air-Drying Treatments

While immersed in water, the fibers of leather are supple and flexible. This works to the advantage of the conservator during treatment since the leather object is relatively easyto work with as compared to leather stored in alcohol or water/alcohol solutions. Often a mixture of alcohol and water are used for the long-term storage of leather since such solutions inhibit the growth of fungi and bacteria. Use of organic solvents, however, displaces water within the matrix of the leather. Loss of water as a lubricant in the leather results in stiffness. In some circumstances, this is useful during treatment since it adds a degree of rigidity to the artifact, preventing damage or loss of information due to disarticulation. The state of deterioration and the need for internal stabilization are prime considerations when deciding on a treatment strategy for leather.

Treating waterlogged leather with PEG followed by freeze-drying to remove water from the artifact has proved to be a highly successful conservation strategy. Because PEG is soluble in water, it is an attractive treatment method for leather. To a lesser degree, PEG is soluble in methanol. Because of this, badly deteriorated leather can be stabilized in a solution of water and alcohol, which helps to give the leather some degree of rigidity. At the Preservation Research Laboratory, treatments usually involve incremental additions of PEG to water to form an aqueous solution in which the leather remains immersed until a high enough percentage of PEG is introduced into the matrix of the leather to allow controlled, slow air-drying.

Lower molecular weight PEGs (PEG 400 to 600) are clear and somewhat viscous at room temperature. These products are ideal for use in room temperature treatment of leather. After reaching 75–85% concentration of PEG in water, the artifact can be removed from the solution, surface-dried of excess solution, and allowed to slowly air-dry in a controlled environment. Controlled drying will prevent excess pooling of PEG on the surfaces of the artifact. Surface deposition of PEG can be removed by daubing the affected areas with a damp, lint-free cloth. Surface pooling, however, may be an indication of more serious problems affecting the stability of the artifact. If the leather is allowed to dry too quickly or if temperature and humidity are not regulated, PEG will migrate to the surfaces of the artifact and, over a short period of time, shrinkage may occur. Controlled, slow air-drying will prevent most of these complications. Generally, however, as much as 10% shrinkage has been associated with the PEG treatment of waterlogged leather.

Freeze-Drying PEG-Treated Artifacts

Generally, there is no steadfast rule for determining the percentage of PEG that should be added to water to pretreat leather before freeze-drying. Experience is the best guide since state of degradation, possible fungal problems, and presence of oxides all affect the potential for preserving leather. Conservators use a wide range of pretreatment strategies in preparation for freeze-drying. In the case of badly deteriorated leather, a 50% or higher aqueous solution of PEG may be necessary to stabilize the artifact. Leather in good condition may require a lower percentage pretreatment.

Aqueous solutions of PEG may require the use of biocides to control biological activity, especially at lower percentages. Because it is impossible to accurately assess the ramifications of their use to control microbial activity, they should be closely monitored and applied sparingly. Many biocides are mildly toxic, and some people have a mild reaction when they come into contact with these agents. Ortho-phenylphenol, commonly known as Dowicide 1, has been widely used since it appears to be an effective deterrent to microbial activity. Tests indicate that it is less of a biological hazard than other commercially available biocides.

After pretreatment, pooled PEG must be carefully removed from the surfaces of the leather before freeze-drying. This will ensure that the freeze-drying process is uniform. All conservators seem to have a freeze-drying strategy that works well for their range of artifacts and levels of expertise. Many artifacts have been successfully preserved using long-term controlled storage in a standard freezer. Humidity control is limited using a conventional freezer system; nonetheless, ice crystals sublimate as gases, resulting in a stabilized artifact. With more elaborate freeze-drying systems, the vacuum, temperature, and relative humidity can all be monitored and regulated. Controlled environment processing and monitoring of humidity allow the conservator to determine when the treatment process is complete.

The use of larger and more elaborate freeze-dryers allows the conservator more control over the processing of artifacts (fig. 4.1). Artifacts are usually frozen for a period of time at very low temperatures (−30°C) prior to placing them into a freeze-dryer. Drying can then be accomplished in many ways. Low-temperature freeze-drying takes a longer period of time, while near room temperature processing will be completed comparatively quickly. Many successes have been reported using ultralow and near room temperature methods, so it is best to experiment with a particular assemblage of equipment to determine how to reliably obtain consistent results. Regardless of which strategy has been prescribed for an assemblage of artifacts, low vacuum is maintained throughout processing.

Other processes have been successful for the treatment of wet leather. J. M. Cronyn reported success in using freeze-drying processing with PEG 400 in 2-methylpropanol.[2] Several nonpolar solvents, such as toluene and kerosene, have also been reported to give good results. Because many of these solvents are flammable, care must be taken in their use. Be aware that changes in pressure and dispersion of fumes will alter the explosion point of some solvents.

In recent years, PEG has fallen into disfavor with many conservators because of changes in environmental laws, which vary from one country to another. Disposal of PEG has become an issue for many laboratories since the introduction of too much

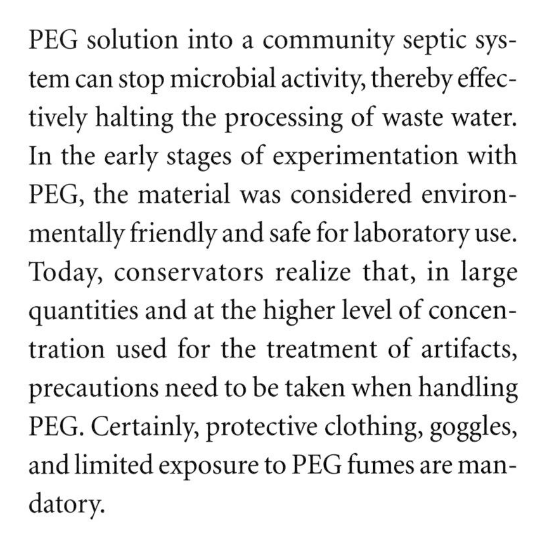

PEG solution into a community septic system can stop microbial activity, thereby effectively halting the processing of waste water. In the early stages of experimentation with PEG, the material was considered environmentally friendly and safe for laboratory use. Today, conservators realize that, in large quantities and at the higher level of concentration used for the treatment of artifacts, precautions need to be taken when handling PEG. Certainly, protective clothing, goggles, and limited exposure to PEG fumes are mandatory.

PEG and Other Polymers

Polyethylene glycol is a polymer. At this point in the history of conservation chemistry, a great deal of experimentation is required to understand the complex relationships among polymers, organic materials, and the chemical reactivity of resins and organic materials. Many processes once thought to be reversible are really not reversible at all. PEG treatments fit into this category. Another example is the treatment of organic materials with acetone rosin. When completed, this treatment essentially encapsulates the entire artifact, resulting in the restructuring of the artifact into what has been described as a single resin molecule. Experimentation, advances in chemistry, and the advantage of several decades of accumulated data suggest that it would be better if conservators dispense with reversibility as a core issue in selecting conservation strategies and, instead, focus on the issue of re-treatability of artifacts.

Initially, many conservators attempted to treat waterlogged leather with high concentration aqueous solutions of PEG. This process has been largely abandoned since, like wood, leather tends to absorb the bulking agent more effectively when applied incrementally. It is possible that high concentrations of PEG do not permit a balanced displacement of water for PEG, resulting in disfigurement of the artifact. Because waterlogged leather is relatively porous, however, almost all molecular weights of PEG can be used.

Fig. 4.1. Modern freeze-dryer used for PEG-freeze-drying large artifacts.

Dry and desiccated leather, on the other hand, may be more receptive to treatment using lower molecular weight materials, which tend to be less viscous. Higher molecular weight PEG 1500 tends to be white in color and has the consistency of paste at room temperature, whereas PEG 3350 remains hard. The elevated temperatures required to melt higher molecular weights of PEG make them more difficult to use with delicate leather.

Conservators have successfully treated thin strips of leather using topical applications of PEG. Many vessels have been spray-treated over long periods of time to facilitate the absorption of PEG into large timbers. Because of their generally smaller size, however, the conservation of leather artifacts is usually accomplished by immersing them in water and using incremental additions of PEG.

Much of the popularity of PEG stems from its ease of use and relatively low cost. Because

PEG dissolves in water, certain considerations must be kept in mind. Artifacts require proper curation to prevent damage. Once treated, they remain hygroscopic and require a temperature- and humidity-controlled environment to remain stable. Over time, the artifact will become acclimatized to a museum environment by either absorbing or giving off moisture to equalize with environmental conditions in the museum. The absorption of humidity from the environment results in a loss or alteration of the original bulking characteristics of the PEG. This, in turn, compromises the stability of the artifact. In some cases, PEG-treated artifacts remain greasy after treatment. Many conservators have avoided this condition by treating leather with blends of PEG, resulting in slightly less supple leather.

It is also important to control exposure of organic materials to ultraviolet light. Over long periods of time, this range of the light spectrum causes damage to leather and acts to chemically alter the nature of the PEG. The chemistry of leather treatment is complex, and because some degree of bonding occurs with the introduction of hydroxil-rich PEG to the leather, this process cannot be completely reversed. This does not preclude the fact that artifacts treated with PEG can often be successfully re-treated.

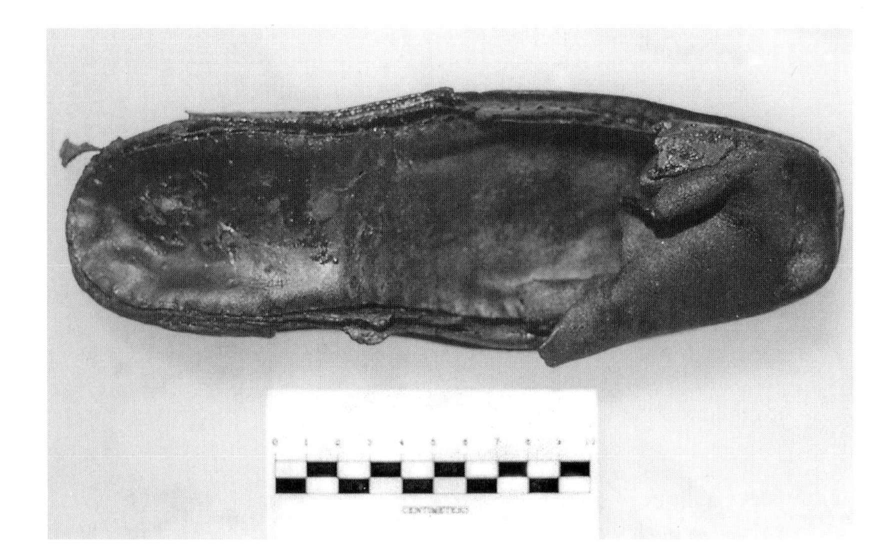

Fig. 4.2. A nearly complete waterlogged shoe.

Passivation Polymer Processes

Passivation Polymers are part of a series of chemistries and techniques developed at Texas A&M University for the purpose of archaeological preservation. These processes were developed in response to the need for artifacts that were more durable and stable after conservation. Passivation Polymers can be used for the treatment of both waterlogged and desiccated leather artifacts. Treatment procedures differ for artifacts recovered from marine sites since it is essential to displace the water before introducing a suitable Passivation Polymer and cross-linker. The case study presented below concerns a leather artifact that was recovered from *La Belle,* which sank off the Texas coast in 1686.

Case Study: A Successful Treatment Strategy for a Waterlogged Shoe

Upon arrival at the Preservation Research Laboratory, one of the shoes recovered from the Texas Historical Commission's excavation of the vessel *La Belle* was placed into freshwater for extensive rinsing. For several months, the shoe was allowed to desalinate along with numerous other artifacts in a vat of freshwater that was changed on a regular basis. Before conservation was begun, the shoe was photographed and thoroughly documented.

The most common methods of preserving waterlogged leather include variations of impregnation of the artifact with PEG. Once completed, the PEG will prevent the structurally deteriorated matrix of the leather from collapsing during the preservation process. While these processes are effective conservation strategies, they are time-consuming and may produce artifacts that require special curation and handling.

PEG dissolves in water, making it a convenient agent to fill the damaged cellular structure and lubricate the fiber bundles of the artifact. Because PEG remains miscible in water after an artifact has been treated, however, long-term stabilization can pose problems. Changes in humidity, temperature, or both can cause the PEG to become unstable. In severe environmental conditions, intercellular movement of PEG within and between the cellular structures of an artifact can cause significant structural damage. Intracellular migration of PEG can result in surface pooling and, eventually, the loss of diagnostic attributes associated with an artifact.

Research at Texas A&M University has culminated in the development of effective new treatment processes for preserving waterlogged and desiccated leather artifacts. These processes are easily implemented using basic equipment commonly found in a modestly equipped conservation facility. Most important, these treatment strategies stabilize objects, requiring little controlled curation.

To demonstrate a basic process of silicone oil preservation, a waterlogged shoe dating from the mid-seventeenth century was preserved (fig. 4.2). An initial inspection of the shoe indicated that much of the upper structure and heel were missing. The insole, sole, and toe portions were in excellent but friable condition. Prior to treatment, there was a slight separation between the insole and sole sections. Much of the stitching, however, was in good condition and appeared to be holding the sections together quite well. Prior to treatment, the shoe was placed into a large container and rinsed in running fresh tap water for several days to remove soluble salts and other debris. Titration testing was used to determine the length of the rinsing treatment.

The shoe was then immersed in fresh acetone. After three days of passive water/acetone exchange at room temperature, it was placed into a new container of fresh acetone. While immersed in this acetone bath, the shoe was placed into a vacuum chamber with a slight vacuum (28 Torr) applied to the artifact to ensure maximum removal of free-flowing water. Throughout this last step of water/acetone exchange (WAE), a piece of aluminum screen was friction-fitted over the shoe, pressing against the sides of the container housing the artifact. This screen ensured that the artifact remained fully immersed in acetone throughout the dehydration process. (See fig. 4.3.)

After the WAE process was complete, the shoe was placed into another container and immediately immersed in a PR-12 Passivation Polymer solution with a 3% addition of CR-20 by weight. As before, a piece of friction-fit aluminum screen was placed over the artifact in solution to ensure that it remained completely immersed throughout the acetone/silicone oil exchange (ASE) process. While in this solution, the shoe was placed into a bell jar vacuum chamber, and at room temperature, a vacuum of 28 Torr pressure was applied. When small bubbles were steadily streaming from the artifact, the vacuum pump was turned off and the artifact was left under pressure for 3 hours. (See fig. 4.4.)

After 3 hours of ASE at pressure, valves were opened, returning the solution to ambient pressure. The artifact was left in solution for 24 hours at room temperature and ambient pressure. It was then removed from

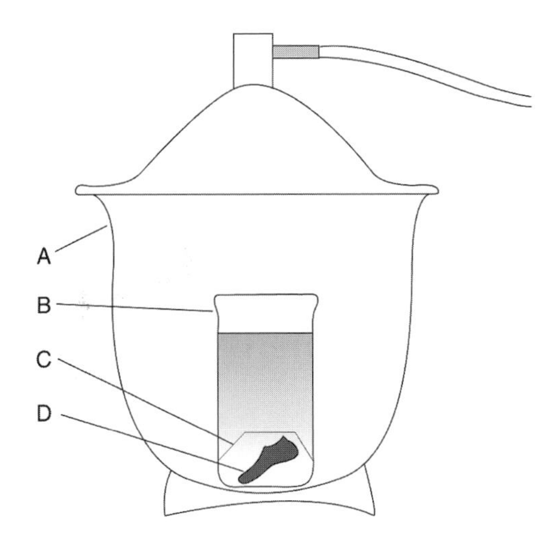

Fig. 4.3. Apparatus used for dehydrating waterlogged leather: (A) vacuum chamber; (B) container with fresh industrial grade acetone; (C) aluminum screen over shoe; (D) shoe immersed in acetone.

Fig. 4.4. Shoe in vacuum chamber, housed in a container with a ventilated cap to prevent splashing while in treatment.

the silicone oil solution and placed on an aluminum screen; free-flowing silicone oil solution was allowed to drain from it. Note that, at this point, exposure to air is no longer a concern and the conservator can take time to situate or arrange the artifact before continuing to the last stage of polymerization.

After allowing the polymer solution to drain from the shoe, lint-free cloths were used to lightly pat the surfaces of the leather. This helped to remove pooled areas on the surfaces. Lint-free cloths with several drops of CT-32 catalyst were then used to wipe the surfaces of the leather. A thin, even coating of catalyst was applied to these surfaces. After approximately 5 minutes, clean cloths were used to wipe any remaining catalyst and silicone oils from the shoe's surfaces. The artifact was lightly polished using dry, soft, lint-free cloths. The shoe was then placed into a large Ziploc bag. A small piece of wadded cloth, dampened with several drops of catalyst, was placed in the bag with the shoe. After the bag was sealed, the artifact was left in close contact with catalyst fumes for an additional 48 hours. Exposing an artifact to catalyst fumes ensures that polymerization will continue into the deep matrix of the leather such as the thick sole and heel.

After treatment, the shoe was removed from the bag and allowed to sit in a well-ventilated fume hood for several hours. Because we were not sure how friable the artifact was during treatment, we did not wipe the surfaces of the leather as thoroughly as we could have. Posttreatment was conducted using wooden dowels and a very soft brush to remove small areas of pooled polymer along the line of stitching on the insole of the shoe. Posttreatment evaluation revealed that no measurable shrinkage occurred as the result of treatment. The diagnostic features of the shoe have been maintained to this day, and the artifact does not require special curation or handling. (See fig. 4.5.)

Accelerated aging tests have indicated that archaeological leather samples preserved using these technologies are stable and not damaged by fluctuations in temperature, humidity, or ultraviolet radiation. Data collected from extensive testing of this specific Passivation Polymer indicate that the artifact will remain unchanged and completely stable for at least 250 years. The polymer will remain supple beyond this time frame and, if needed, the artifact can be re-treated.

Fig. 4.5. Bottom view of a nearly complete shoe.

Passivation Polymer Treatment for Desiccated Leather

Dry and desiccated leather artifacts are often fragile, making excessive handling a problem. In many respects, waterlogged leather is easier to work with since water acts to lubricate the collagen fibers. Leather recovered from cave sites and other dry environments is often more difficult to restore since the lack of moisture causes collagen fibers to shrink. Once this has occurred, it is sometimes impossible to soften the leather and reshape it. Dry and desiccated artifacts, nonetheless, require conservation to prevent further deterioration.

Desiccation and dry salty conditions may actually help preserve some forms of leather. Cronyn suggests that rawhide, tanned and semitanned materials, and parchment may be better preserved due to desiccation or close association with a salty environment.[3] Salt, he points out, acts to desiccate leather and inhibit damage caused by microorganisms. In many situations, disintegration of the leather may be advanced to the point that it must be consolidated to prevent further loss of surface details. Polymers have proved to be an effective agent for preservation of dry leather and for the consolidation of leather that has become fragile over time. In many cases, the wide variety of Passivation Polymers being used for artifact conservation enables the conservator to select polymers with sufficiently long structures that will increase the flexibility of dry, stiff artifacts.

Unlike artifacts recovered from marine environments, much of the leather from terrestrial sites does not need dehydration prior to treatment with polymers. Leather recovered from these sites, however, usually needs surface-cleaning. Often gentle wiping with a lint-free, damp cloth will remove encrusted surface dirt. Fortunately, leather from terrestrial sites is seldom encased in harder concreted materials. When slight encrustation is present, soft wooden dowels or dental tools may be required to clean these areas. Stains are a common problem associated with leather artifacts from land sites due in part to their close association with other artifacts, absorption of minerals in the soil, and bacterial and fungal activity. All these problems must be addressed to ensure long-term stabilization.

Flat pieces of dry leather can be treated using minimal equipment; and since it is not necessary to dehydrate the leather before continuing treatment, processing can be accomplished in a short period of time. Because dry leather is often brittle, it is helpful to place the artifact between two sheets of glass to prevent stresses caused by handling. Small elastic bands or binder clips can be used to sandwich the leather between sheets of glass (fig. 4.6).

Leather shoes and artifacts that are not inherently flat cannot be placed between sheets of glass prior to treatment. A number of precautions can be taken, however, to ensure that these artifacts will be correctly shaped and aesthetically accurate. Loosely crumpled gauze placed inside a shoe works well to maintain its shape during treatment. The presence of such an open-structured material does not affect the ability of a Passivation Polymer to penetrate into the leather. Expandable polyethylene mesh bags, similar to those used in the produce section of a supermarket to hold fruits and vegetables, are an ideal way to keep small and fragmented artifacts together during treatment. Because of their open-weave design and inherent flexibility, these bags have proved to be extremely useful in handling fragile artifacts.

Water/Acetone Exchange

Prior to treatment with polymers, water must be removed from the leather. To prevent damage, the leather must be slowly dehydrated, starting first with a series of ethanol baths and working up to at least two baths of acetone. This will appreciably reduce shrinkage and problems of hardening related to the locking of fibers as water is displaced within the leather's matrix. While immersed in water, waterlogged leather remains supple. The initial stages of treatment are therefore the most opportune times to adjust and shape the artifact. In successive baths of ethanol and acetone, the artifact will become more rigid. Once the artifact has been dehydrated in acetone, it should remain in the final bath until a solution of polymers is prepared. *Removing it from acetone and allowing it to quickly air-dry will cause irreversible damage.* Dry and desiccated leather, on the other hand, needs only to be surface-cleaned prior to treatment in polymers.

Choosing a Suitable Polymer or Combination of Polymers

Polymer technology has progressed to the point that conservators have a wide range of viscosities and controllable attributes to choose from for the treatment of artifacts. Just like PEG, polymers can be mixed in infinite combinations to allow the conservator control over such attributes as flexibility, tensile strength, texture, and degree of consolidation that will be imparted to an artifact.

At present, acetone/polymer exchange is the mechanism most commonly used to impregnate an organic substance with polymer/cross-linker solutions. Normally, acetone in the matrix of an artifact will evaporate quickly at room temperature and ambient pressure. Unless the artifact is immersed in a suitable replacement media, the acetone will vaporize and, with nothing to displace the acetone, shrinkage and distortion will occur. In a vacuum, the boiling point of acetone is reduced, resulting in faster evaporation. Controlling the rate of evaporation and the rate of displacement of acetone with a suitable bulking agent, then, is the key to successfully maintaining the diagnostic attributes of an artifact.

Rate of evaporation of acetone is only one factor that can be controlled. Viscosity, the length of the polymer "backbone" or centistoke of polymers, and the temperature at which acetone/polymer exchange is conducted are important in controlling the end product. Generally, shorter chain polymers, such as PR-10 and PR-12, are easily exchanged into the matrix of leather because of two factors. First, their relatively small size makes it easier for them to penetrate into the open and porous structure of waterlogged leather. Second, because of their small size, these polymers tend to permeate more quickly than larger centistoke polymers. Leather treated with smaller centistoke polymers tends to be more rigid than leather treated with larger centistoke polymers.

PR-14 has a relatively long chemical backbone compared to PR-10 and PR-12. Artifacts treated with PR-14 tend to be more supple after treatment. By blending polymers, the conservator can design a treatment strategy specifically geared to treating problem aspects of an individual artifact. For example, in a badly deteriorated, waterlogged shoe, blends can be mixed to ensure that the surface integrity of the leather is preserved as well as the thicker, more stable areas of the artifact.

An Effective Treatment for Dry Leather

Dry leather does not require dehydration in acetone before treatment. If the artifact has been surface-cleaned using water, air-drying in a well-ventilated fume hood for at least 24 hours is necessary. The movement of air will ensure that sufficient moisture has been removed. If the artifact is being treated between sheets of glass, the time required to ensure dryness will be longer.

Once the artifact is cleaned and prepared for treatment, it can be carefully immersed into one of the PR-series polymers to which CR-20 has been added in the amount of 3%

by weight. Normally, it is best to friction-fit a piece of aluminum screen over the top of the artifact in solution to ensure that it remains immersed throughout treatment. Because acetone is not being used as the driving mechanism to transfer polymers into the leather, a slight vacuum can assist the process. If the leather is too fragile or deemed to be unsuitable for withstanding vacuum pressure, the process of impregnating the artifact can be accomplished at room temperature and ambient pressure. This process, naturally, will take longer.

For sections of leather ranging from 2 to 6 mm thick, it takes approximately 20 minutes for the polymer solution to be adequately absorbed. The use of vacuum will speed up absorption. The process must be closely monitored, however, because if air is drawn from the artifact faster than polymers can fill the matrix of the leather, some distortion may occur. To avoid these problems, use vacuum cautiously. Over a short period of time, all the PR-series polymers will successfully penetrate dry and desiccated leather.

After treatment in a polymer solution, the artifact should be allowed to drain for at least 2 hours. Because the polymer solution can be reused several times, place the artifact on aluminum mesh screen over a large container, so that runoff polymers can be recovered. During this phase of treatment, it is wise to spend as much time as needed to position and pose the artifact and to ensure that everything is done to make the leather natural looking. Using soft lint-free cloths or gauze to gently daub the surfaces of the artifact will remove pooled polymer solution and prepare the surface for catalyzation. Once catalyst has been applied to the leather's surfaces, cleanup will be a little more labor intensive. In situations where the leather is very fragmented or fragile, it may be better to do minimal cleanup before catalyzation with the intention of working more with the artifact after it has been stabilized and is capable of being handled.

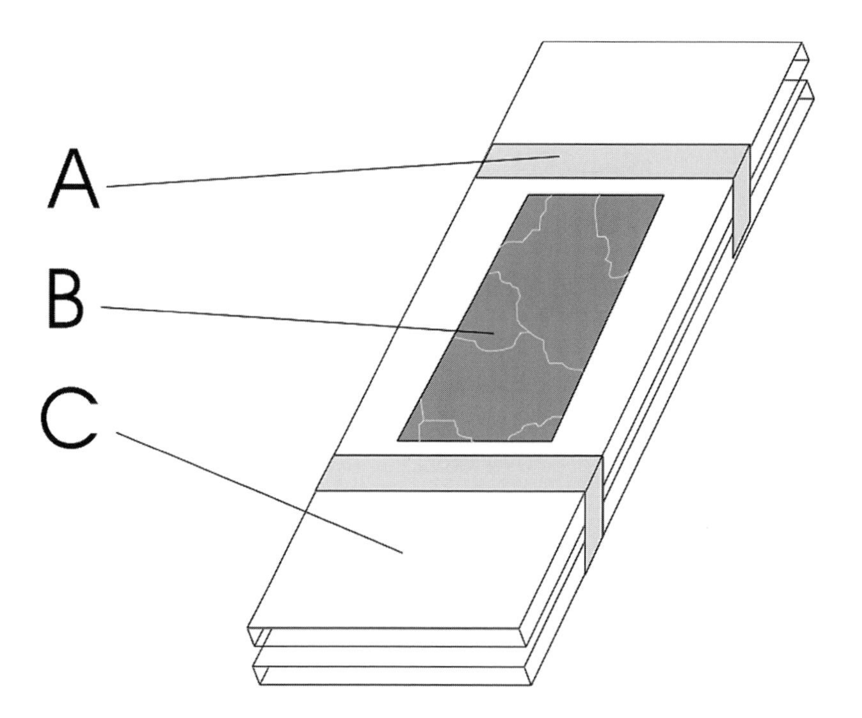

Fig. 4.6. Glass plates used to support fragile leather during treatment with silicone oils: (A) rubber bands; (B) leather; (C) clear glass plates.

Once the artifact is aesthetically correct and free of pooled polymers on its surfaces, catalyst can be applied. This phase of treatment should be conducted in a well-ventilated area, preferably in a fume hood. Transient vapors may contaminate open vats of polymer/cross-linker solutions that are in close proximity to the workspace. The conservator must also avoid personal exposure for health and safety reasons.

CT-32 is a good catalyst for leather artifacts because it can easily be applied to the surfaces of the leather using soft cloths or cotton swabs. Regardless of the means of application, the artifact should be covered in a thin, even layer of catalyst. After 4 to 5 minutes, use a soft cloth to remove the catalyst and any polymer solution that may have pooled on the surface of the artifact as a result of chemical reactivity with the catalyst. Successive soft cloths can be used to gently wipe the surfaces of the artifact. In a short period of time, the leather will have a natural texture and feel. Even though most of the catalyst has been removed from the surfaces of the artifact, a chemical reaction with the polymer/cross-linker-rich matrix of the leather will continue for approximately 24 hours. After topically applying catalyst to leather, it is often advisable to place it into an airtight environment with a small dish of CT-32 and allow the catalyst vapors to continue the process of catalyzation. The process is the same as that suggested for delicate artifacts that cannot withstand handling.

In situations where the leather is too fragile to withstand surface-wiping, catalyzation can be accomplished using vapor deposition. The simplest means of initiating catalyzation is to place the artifact and a small dish containing a few grams of CT-32 into an airtight environment and allow the artifact to remain in treatment for at least 24 hours. After 24 hours, the dish of catalyst will need to be replaced with a fresh one. To increase the volume of vapors, the containment chamber can be placed into a ventilated warming oven set at 37°C. If additional treatment is deemed necessary, new catalyst needs to be placed into the containment chamber every 18–24 hours.

After vapor deposition, all leather should be placed in a well-ventilated area for at least 72 hours. This will ensure that the residual smell of catalyst disappears. Exposure to air will also ensure that the polymer is set and the artifact is completely stable.

Suggestions for Treating Leather between Sheets of Glass

Glass sheets make it easy to work with fragile leather and, with a few precautions, the process is highly successful. Because the surfaces of the leather are obstructed by the glass sheets, the process of acetone/polymer exchange will be slower; the use of vacuum will speed up the process considerably. Once polymers have been introduced into the leather, the artifact can be removed from the polymer solution and allowed to drain of free-flowing solution. This process, too, will be slower since the viscosity of the polymer solution and the available free space between the sheets of glass will work to slow drainage.

Once sufficiently drained, the elastic bands or binder clips holding the artifact in place should be removed and the top glass plate gently slid away. Blotter material or soft gauze can then be used to gently daub the surfaces of the leather and surrounding surfaces of glass. Once the top surface is dry, a clean glass plate should be placed over the top surface of the artifact and the entire artifact turned upside down so that the bottom surface of the artifact and covering glass plate are on top. This glass is then removed and the process of cleaning the bottom surface of the artifact is completed. Once cleaned of free-flowing polymers, the artifact should be transferred so that it is sandwiched between thin sheets of lint-free cloth between two sheets of clean glass. After rubber bands are applied to hold the encased artifact together, vapor catalyst deposition can be completed. The process sounds involved but is actually quickly accomplished with minimal stress to the artifact.

Storage and Display of Leather Artifacts

The excavation, documentation, recovery, and conservation of artifacts are costly and time-consuming. All the labor and research invested in an assemblage can be lost in a relatively short period of time if the artifacts are not properly stored and displayed. It is, therefore, important to have a detailed plan for the long-term storage of artifacts before treatment is initiated. Chromium tanned leather is, in part, more resilient to some atmospheric pollutants than vegetable tanned leather. But other factors play a role in the safe storage of leather. Treatment methods are also important. Because of the water miscible nature of PEG, artifacts treated using this method require a more controlled environment than leather treated with Passivation Polymers. Other considerations are relative humidity, exposure to natural light, microorganisms, airborne pollutants, and handling by researchers and museum personnel.

Relative Humidity

Most organic artifacts are hygroscopic. Waterlogging processes act to break down the fibrous structures of leather as well as remove natural lubricants that protect fibers from hydrolysis. Hydrolysis increases the amount of water in the matrix of the leather, which ultimately must be replaced with an agent that can stabilize the leather without its losing physical integrity. Relative humidity is an important consideration for the storage of leather since elevations in humidity can promote microbial activity and new problems in the stabilization of the artifact.

Maintaining a constant relative humidity in a display environment will prevent absorption and release of moisture from leather. In many cases, leather treated with PEG can swell if relative humidity is too high (70% or more). The miscibility of PEG-treated leather encourages beading and surface pooling of the bulking agent as moisture in the surrounding environment is absorbed into the leather. Ideally, maintaining a constant relative humidity is best for artifacts, and it is a costly environmental factor to ignore. RH is computed using the following simple equation:

$$\frac{\text{Amount of } H_2O \text{ in air}}{\text{Maximum } H_2O \text{ air can hold at a specific temperature}} \times 100 = \text{Relative Humidity}$$

Temperature

Many chemical reactions are slowed or inhibited by keeping organic materials at lower temperatures. Temperatures must never be so low, however, that changes in humidity cause problems with condensation. It is important to keep artifacts at a constant, cool temperature below the melting point of the bulking agents and consolidants used to treat the assemblage. PEG-treated artifacts are most vulnerable to temperature changes because as temperatures are decreased, the relative humidity increases, causing potential problems in the stability of the artifact. Artifacts preserved using Passivation Polymer technologies are substantially more stable and are unaffected by changes in temperature. Although stability is increased, however, good practice dictates that all artifacts be treated with the same concerns for safety and stability.

Light

Exposure to light is one of the most potentially dangerous problems for artifacts in museums. Ultraviolet light causes oxidation and sets a number of chemical reactions in motion that are not fully understood. Many of the consolidants and adhesives used in the preservation of leather are sensitive to light and, over a short period of time, become brittle or lose their desired characteristics.

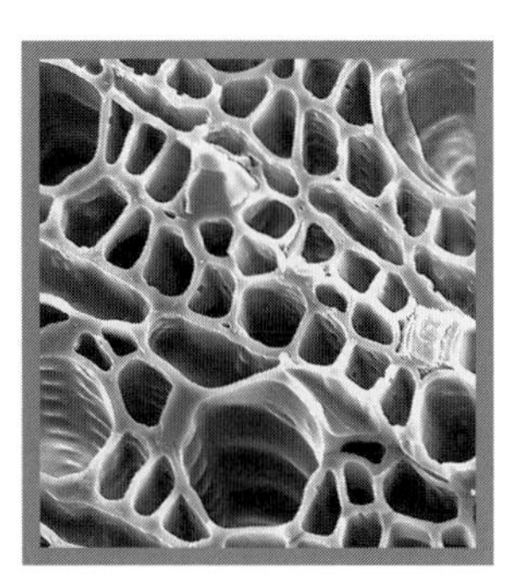

Composite Artifacts

Composite artifacts are generally made of two or more contrasting materials. Examples include iron cannon balls outfitted with iron straps and a wooden base called a sabot, and knives recovered from the shipwreck *La Belle* consisting of hard wooden handles that contain sections of iron knife blade.

Often each component in a composite artifact requires a different treatment strategy. If this is the case, disassembling the artifact into component parts may be essential for preservation. Two complications arise when disassembly is required. Waterlogged artifacts are usually structurally compromised. Even the most precise job of disassembly may result in breakage or damage to one of the components. Also, when components are treated separately, they are exposed to treatment-related structural changes that may inhibit posttreatment reassembly. Wooden components, for instance, are more prone to swelling during preservation. During treatment, badly deteriorated iron may be prone to surface spalling, resulting in the loss of metal. Even the most gifted conservator will encounter problems in treating some composite artifacts.

One cannot overstress the point that silicone oils preservation techniques are not a cure-all for the conservation of all artifacts. These technologies, however, do offer some benefits for the preservation of delicate artifacts concreted together with heavy iron artifacts. Such is the case of a waterlogged wicker basket, heavily concreted with iron oxides and iron shot, which was recovered from *La Belle.*

Case Study: Preservation of a Composite Artifact Containing Basketry and Iron Shot

The silty bottom of Matagorda Bay has created the ideal anaerobic environment in which the large assemblage of artifacts from *La Belle* have been well preserved. When recovered, the wicker basket and iron shot were indistinguishable, since oxides from the cast iron balls had combined with minerals and sediments to create a large formless block, commonly referred to as a concretion. Apart from the general outline of a few of the larger shot, the only distinguishable feature was a large wooden barrel stave encrusted to the flattest surface of the concretion.

Once the stave and layers of concreted material were removed, the flattened outer surface of a wicker basket was visible. Additional removal of encrustation revealed that a bar shot was also concreted to the top surface of the block. Before an initial evaluation could be made, a pneumatic chisel was used to remove the bar shot from the outer surfaces of the artifact. Radiographs and mechanical cleaning of a small area of the concretion's flattest surface indicated the outline of one side and the base of a basket. From the X-rays, the basket appeared to be intact but heavily permeated with oxides and concreted material.

Inside the basket, 32 small shot were clustered, held in place by impacted sediments and concreted materials (fig. 5.1). The shot ranged in size from 4.8 to 5.0 cm in diameter. Because the outer surface of the basket appeared to be intact, the decision was made to clean the thin layer of encrustation encasing this outer surface. To minimize damage to the fragile wicker basketry, a rubber cast of the bottom was made. During treatment, this cast would act to support and protect the basketry while the shot were mechanically cleaned and removed from the interior of the basket.

Materials

Because of the weight of the overlying concreted cast iron shot, high tensional strength and dimensional stability were two criteria used in selecting molding materials. HSIII RTV silicone elastomer compound (25,000.00 MPA.S) was chosen. This silicone rubber is ideal for capturing surface details. It is off-white in color and not prone to hazardous or unplanned polymerization due to chemical reactivity with oxides or other materials. HSIII RTV has been widely used for restoration castings because of its ability to capture fine surface details and its long-term stability and ease of use.

The weight of the shot and the fragility of the wicker necessitated the creation of a stable, solid support for the basket that would not cause stains to the organic component of the artifact or cause additional cleaning problems in the crevices of the basketry or iron shot. During initial testing, HSIII RTV did not react with other resins and casting materials that might potentially be needed as treatment progressed. Significantly, this casting material did not interact adversely with the wet surfaces of the wicker and iron.

Concerns have been raised that staining can occur to porous surfaces, due to the release of very low molecular weight silicone compounds during setting.[1] In creating a successful facing compound for the treatment of painted plasters, Pulga concluded there was a direct link between the amount of staining that occurred and the porosity of the organic material being treated.[2] Pulga and others have noted that shorter setting times reduce the release time of siliceous compounds and, therefore, reduce the risk of staining. Since the basketry was thoroughly stained with minerals and surface concretion, potential staining caused by siliceous compounds during treatment was determined to be minimal.

Displacement Mechanism and Chemical Sequence

In order to introduce a polymer/cross-linker solution into the basketry, it was first necessary to eliminate water in the wicker and surrounding concretion and iron. Baths of industrial-grade acetone were used to thoroughly dehydrate the concreted artifact. Because of its volatility, acetone could easily permeate thin layers of encrustation to ensure that all surfaces of the basket were dehydrated prior to the application of the polymer. After dehydration, a polymer/cross-linker solution could be introduced into the matrix of the basketry by a process of passive displacement of evaporating acetone.

For this project, PR-10, a hydroxyl-ended, functional silanol was combined with Q9-1315 cross-linker (20% by weight) to form a low viscosity polymer capable of displacing

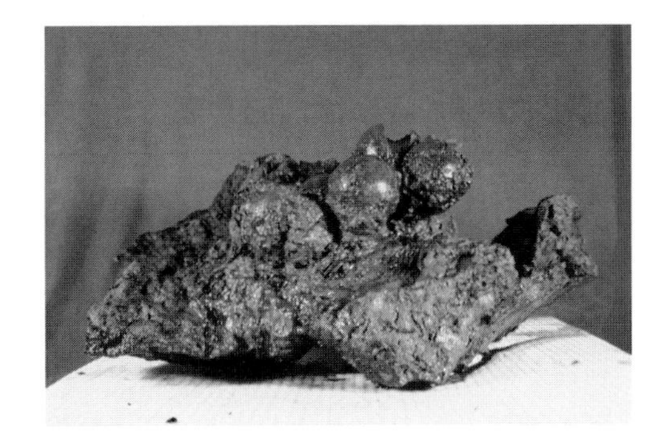

Fig. 5.1. Concretion with shot visible on upper surface.

acetone in the matrix of the basketry. Q9-1315 is a hydrolyzable, multifunctional silane capable of tying two or more polymer chains together. CR-20 and Q9-1315 cross-linker, which is methyltrimethoxysilane diluted with methanol, are similar in structure to tetramethyloxysilane, with one of the methoxy functional groups replaced with a methyl group. The methyl group increases the electron density of the silicon atom, causing the reaction to be completed more quickly.[3]

After acetone was displaced with the polymer solution, the basketry was cured using a topical application of a dibutyltin diacetate catalyst followed by several days of vapor catalyst deposition in a controlled environment. The mechanism of preservation is a combination of surface consolidation and thorough penetration and polymerization of polymers throughout the basketry.

Chemical Reaction

Many variations have been offered regarding the mechanisms involved in the process of polymerization. The basic mechanisms, however, are not disputed. Alkoxysilanes undergo a process of hydrolysis/condensation to form a cured, cross-linked matrix. The alkoxy groups are first hydrolyzed with atmospheric water, evolving an alcohol and thus forming silanol groups (fig. 5.2, *1*). This results in condensation of the silanol groups, either by alcohol or water condensation. This reaction can be extrapolated to explain the curing of a hydroxy-terminated polydimethylsiloxane using alkoxysilane as a cross-linker.[4]

While there are differing theories about the mechanism, the most probable process is the cleavage of the Sn-O-Si bond, which links the alkoxysilane with the polydimethylsiloxane (fig. 5.2, *3*). The reaction continues with the remaining methoxy and hydroxy groups until all have reacted or until the resulting cross-linked polymer is too fixed for the remaining methoxy groups to find remaining hydroxy groups. In this reaction, alkoxysilane is considered to be a cross-linker because it cross-links the linear polydimethylsiloxane into a three-dimensional matrix.

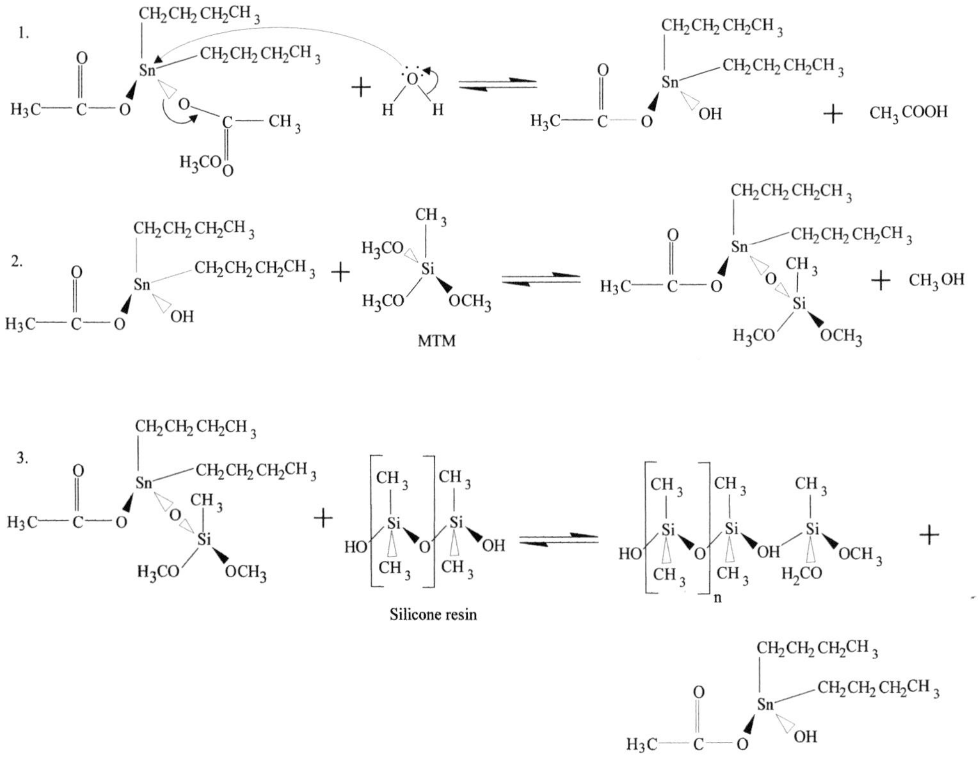

Fig. 5.2. Formula depicting the process of polymerization.

Methods—Removal of Concreted Material and Initial Treatment of the Basket

Prior to treatment, extensive radiographs of the artifact were taken. Traces of the outline of the basketry and shot were transferred onto Mylar. These drawings were necessary to assist in the removal of the shot with minimal disturbance to the underlying basketry.

The layer of concretion covering the base of the basket was very thin (0.91mm) due in part to a barrel stave covering a section of the base. A pneumatic chisel was used to remove the stave and the underlying layer of calcareous material. Dental tools were used to remove some of the calcareous material between the strands of wicker. Care was taken to ensure that the basket remained wet throughout this phase of mechanical cleaning. The entire concretion was then placed into a large vat of industrial-grade acetone. After 36 hours of room temperature dehydration, it was removed and rapidly transferred into a vat containing a silicone oil solution consisting of PR-10 silicone oil to which a 20% addition of Q9-1315 cross-linker had been added, by weight, and thoroughly mixed. The artifact was allowed to sit in the polymer solution for seven days. The artifact was then removed from the solution and placed onto several layers of newspaper, where it was allowed to drain of free-flowing polymers. Soft cloths were used to wipe additional polymer solution from the surfaces of the concretion and from the irregular surfaces and recesses of the basket.

Molding the Bottom of the Basket

Once the thin layer of concretion covering the flattest surface of the basket had been cleaned, dehydrated, and impregnated with a polymer/cross-linker solution, a cast of the surface of the basket was made, in order to create a form-fitted, supportive base that would protect the exposed section of basket while the upper surfaces of the artifact were cleaned and the individual shot removed. Modeling clay was used to create a deep-sided dam around the exposed edge of the basket. A mold of the bottom surface of the basket was made using HSIII RTV silicone elastomer compound. After the cast was allowed to cure for 24 hours, the clay dam was removed and the entire artifact was turned over so that it was resting in the newly created base (fig. 5.3). The HSIII RTV rubber mold was firm enough to support the exposed surfaces of the basket throughout treatment but soft enough to dissipate potentially damaging vibrations from pneumatic chiseling and manual removal of the iron shot.

Fig. 5.3. Making the HSIII RTV rubber base.

Removing the Shot

Before removing the shot from the interior of the basket, topical applications of the PR-10/Q9-1315 were applied as iron shot were taken out and new areas of the inner surface of the basket were exposed. This ensured that these areas of the basket were impregnated with the polymer solution. Rate of absorption indicated the degree of penetration of the polymer solution into the fabric of the basket. The presence of pooled polymers in areas topically treated with the polymer solution indicated when absorption of the polymer into the basket was complete. Dental tools and pneumatic chisels were then used to remove impacted sediments and oxides from the basket's interior (fig. 5.4).

Spot applications of the polymer solution continued as the remaining shot were removed. Once the internal and external surfaces of the basket were cleaned, the focus of work shifted to additional cleaning of impacted silt and oxide-laden concreted material within the woven recesses of the wicker strands. Dental tools were the safest and most effective tools for this work. Because the wicker was impregnated with the polymer solution, most of the remaining concreted material was easily removed. After more concretion was manually removed, the basket was immersed in a vat containing PR-10/Q9-1315 solution for approximately 8 hours. It was then removed from the polymer solution and placed on several layers of newspaper. Soft cloths were used to wipe the surfaces of the basket. Because the basketry was fully impregnated with the polymer solution, there was no concern that the wicker would dry out or shrink. Accordingly, the basket was allowed to drain of free-flowing silicone oil solution for several hours (fig. 5.5). While the solution was draining, the basket surfaces were wiped several times to remove pooled polymers.

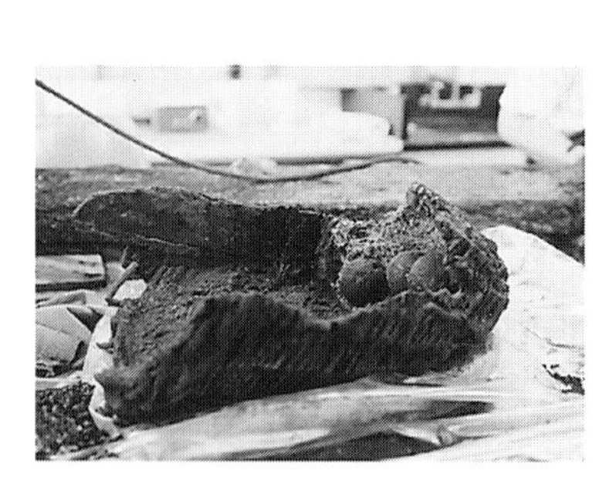

Fig. 5.4. Removal of shot from the interior of the basket.

Catalyzation

CT-32 was used to polymerize the silicone oil impregnated wicker. CT-32 can be applied either topically or as a vapor. Because of the complexities of this project, both methods were used to initiate polymerization. Fragile areas located at the upper rim of the basket were spot treated with CT-32 using Q-tips and soft cloths. This was necessary to stabilize these areas while additional cleaning of the wicker was carried out. After five minutes of chemical reactivity, areas applied with spot treatments of CT-32 were thoroughly wiped with soft cloths.

Once the basket had been cleaned, it was placed into a large polyethylene bag so that the entire artifact could be vapor treated with CT-32 catalyst. Several lint-free cloths containing drops of CT-32 were wadded and placed into the bag along with the basket. When sealed, the polyethylene bag formed an ideal containment chamber, allowing catalyst fumes to remain in close proximity to the surfaces of the basket. After 24 hours of exposure to catalyst vapors, the basket was removed. Soft cloths dampened with Q9-1315 cross-linker were used to remove small areas of surface-pooled polymers from the basket. After a brief cleanup, the basket was returned to the polyethylene bag along with three wadded cloths, each containing several drops of CT-32.

After a total of 48 hours of catalyzation, the basket was removed from its containment chamber and placed into a fume hood. During the polymerization process, a few small areas of white particulate were noted on the surface of the basket in areas that had been spot treated with CT-32. This particulate was easily removed by wiping the affected areas with a lint-free cloth dampened with Q9-1315 cross-linker.

Iron Shot

All the iron shot were conserved using electrolytic reduction. Each shot was placed into a mild steel vat. A form-fitted mild steel anode was constructed using 16-gauge expanded mild steel mesh with ½-inch openings. An all-steel construction C-clamp was used to hold the shot during treatment. A section of 16 AWG, separation 2,300V maximum rating insulated copper wire and Mueller zinc-plated alligator (number 25) were used to connect the artifact to a regulated DC power supply. An aqueous solution of 5% sodium hydroxide solution was used as an electrolyte solution.

Mercuric nitrate testing was used to determine quantitative amounts of chlorides in the electrolyte solution. Once salts were minimized in each shot, they were placed into baths of deionized water and boiled, 8 hours per day, for three days. Each shot was then coated with a solution of Baker's tannic acid, dissolved in ethanol. Three separate coatings of tannic acid were applied to each shot, allowing a ferric tannate barrier to form on the metal's surface. Each shot was then treated in microcrystalline wax.

Discussion

HSIII RTV casting materials worked well for creating a cast of the surface detail of the side

Fig. 5.5. Polymer solution draining from the surfaces of the basket.

of the basket. No chemical reactivity was noted with the wet surfaces of the basket during the casting process and, in spite of the viscosity of the material, highly detailed casts of the basket were obtained. The casting material reproduced undercuts in the basket's surface due to the irregular shapes of the strands of wicker and the variation in the voids between the strands. Because HSIII RTV remains supple, casting of undercuts and irregularities in the outer surface of the basket rendered an extremely accurate reproduction of the basket's surface detail.

Numerous silicone oil/cross-linker combinations were considered for treatment of the wicker. Because of concerns about the inability of a viscous polymer compound to penetrate through oxides and surrounding concretion, Q9-1315 was selected as a cross-linking agent. Q9-1315 contains a high percentage of methanol, which acts to make the resultant polymer solution less viscous. The use of methanol-based materials, however, requires that work be conducted in a well-ventilated fume hood and that appropriate apparel be worn to minimize exposure.

Because the concretion had been thoroughly dehydrated in baths of acetone, displacement of acetone with the polymer solution was effective. As individual shot were removed from the interior of the basket, the use of a topical application of the polymer solution ensured that each newly exposed section of basket was thoroughly treated.

The strategy of casting the partially exposed flat surface of the basket worked well. As the surrounding concretion was mechanically removed, the rubber mold of the surface and its voids and undercuts acted to ensure that the physical size and shape of the basket were being preserved as treatment continued. Because of the friable state of the wicker, removal from the concretion prior to treatment would have compromised dimensional data and other physical attributes of the basket.

The task of cleaning compacted sediments from between the fibers of the basket was a time-consuming task. Because the individual fibers were treated with a silicone oil/cross-linker solution, the process of cleaning was facilitated since the polymer solution acted as a lubricant between the surfaces of the wicker strands and the surrounding sediments.

CT-32 is a versatile catalyst. Because there were numerous delicate strands of wicker around the upper edge of the basket, spot treatment using CT-32 allowed conservators to preserve small areas that were in jeopardy of being lost or destroyed, while allowing the more robust strands of wicker to remain unpolymerized until the entire basket was ready for catalyzation. Acids and bases should be avoided for purposes of artifact conservation; therefore, organo-tin catalysts are desirable since they tend to rapidly oxidize into a neutral substance when exposed to fresh air. Organo-tin compounds undergo hydrolysis with atmospheric water, producing organo-tin hydroxide. These oxides are the agents of catalysis.

For purposes of treating the cast iron shot, electrolytic reduction processing ensured removal of soluble salts from the iron and provided excellent surface-cleaning capabilities. Use of tannic acid followed by treatment with microcrystalline wax has a proven track record for treatment of iron artifacts.

The major benefit of using silicone oils in the preservation of composite artifacts is that the polymer/cross-linker does not have a negative effect on the iron artifact component of the concretion. This allows the conservator to treat the organic component of the artifact before commencing the task of removing the bulk of the concretion. The presence of the polymer solution in the matrix of the organic component prevents it from dehydrating. Polymer-impregnated silts and concretion can usually be easily cleaned from the surfaces of the organic component. This is advantageous for small, fragile sections of basketry.

Spot treatment catalyzation of the thin strands of wicker forming the edge of the basket is advisable when working with larger, cumbersome artifacts. The ability to localize treatment ensures that frail strands are well preserved while work continues on other areas of the artifact.

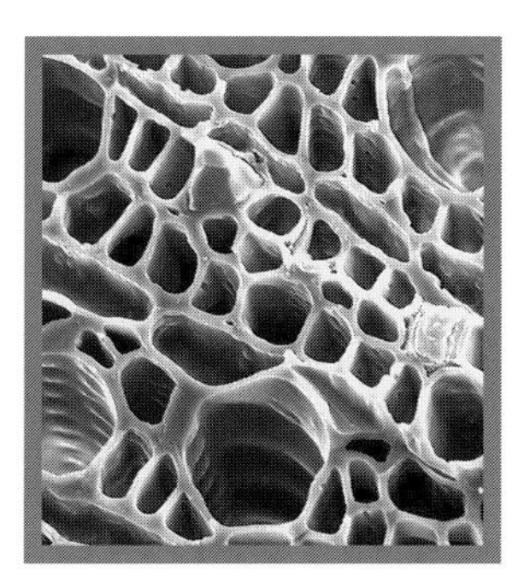

CHAPTER 6

Cordage and Textiles

Throughout history, cordage in one form or another has been made from the natural fibers of plants and animals. Cotton, flax, jute, hemp, grasses, silk, wool, hair, and strips of animal hides have been used to make rope. These materials have been woven, braided, or spun together to create threads for sewing and weaving.

Artifacts made of cotton and flax are greatly susceptible to fungi and bacteria in humid conditions and are seldom recovered from archaeological excavations. Other artifacts made using animal fibers, which are primarily proteins, are better able to survive because they are less prone to destruction by bacteria. All cordage and textiles, however, are affected by exposure to light. Natural dyes will fade quickly if an artifact is left in direct, bright sunlight. The ultraviolet spectrum of normal light will accelerate deterioration of the artifact, often reducing its pliability. Airborne pollutants interact with dyes and natural fibers. Care should be taken to ensure that artifacts are not stored in modern cabinetry constructed using particleboard and urea-based finishes. Off-gases from these materials interact with the natural fibers of the artifact and some chemicals commonly used in their preservation.

Other forms of air pollution can affect museum displays and long-term curation of artifacts. Fumes from automobile emissions are dangerous to organic artifacts. Hydrocarbons, nitrogen oxides (NO_X), and carbon monoxide (CO) greatly affect cordage and textiles. Storage and display of artifacts in close proximity to parking garages should therefore be discouraged. Airborne soot created from incomplete combustion of fossil fuels may chemically react with fabrics. Particulate contaminants lodge between fibers, causing wear and physical damage.

Singly, or in combination, these factors and ever-present microorganisms act to undermine an artifact's tensile strength and pliability. Generally, textile conservation should be left in the hands of specially trained conservators. Accordingly, most of this chapter is devoted to the preservation of waterlogged rope. One case study, however, discusses the preservation of tar-coated canvas associated with a gudgeon strap from a vessel that sank during the earthquake of 1692 in Kingston Harbor, Jamaica.

New Techniques for the Preservation of Waterlogged Rope

To help illustrate new techniques of preserving waterlogged rope, I discuss rope recovered from La Salle's vessel, *La Belle,* which sank off the Texas coast in 1686. Among the organic materials found on the vessel were lines from the rigging and coils of rope that had been stored in her hold. Conservators have been faced with the formidable task of preserving the large amount of hemp rope recovered from the vessel, which includes segments more than 30.5 m long. Especially challenging to the conservators is the anchor line of the vessel, consisting of a continuous section of rope more than 152 m long. The diameter of these sections of rope ranges from 6.25 to 6.6 cm, after conservation. Although most of the recovered rope appeared to be in pristine condition during excavation, microscopic analysis indicated that most of the fibers were thin and visibly degraded; microbial action had caused the fibers to decay while significant water saturation had further weakened them.

Silicone Treatment Strategies

The Texas A&M Archaeological Preservation Research Laboratory technique for stabilizing waterlogged rope with silicone oils involves a displacement of the water trapped in the rope fibers with acetone, followed by the replacement of the acetone with a hydroxyl-ended functional polymer and cross-linker. The polymer-impregnated rope is then cured by exposure to a catalyst, which is applied either topically or as a vapor. I suggest that the preservation of the treated rope is a result of both surface consolidation and penetration of hydroxyl-ended polymers with cross-linking agents added. The polymer-rich matrix of the artifact is then treated with a tin-based catalyst, which completes the polymerization process.

Frankfurter Method of Rope Preservation

Conservators at the National Museum Conservation Laboratories in Brede, Denmark, routinely use a technique for conserving waterlogged rope that they refer to as the "Frankfurter method."[1] This process involves encapsulating waterlogged rope between sheets of perforated polypropylene film; the film is then heat-sealed to produce a formfitting jacket in which each rope sample remains throughout treatment. The packaged rope is attached to a piece of Masonite, which acts to support the rope, and then treated with PEG. After treatment, the rope is placed into a large freeze-drying unit and freeze-dried at –20°C with a 50% relative humidity.

After freeze-drying, the rope sample is removed from the Masonite/polyethylene bag structure and allowed to sit in fresh air. Rope specimens treated with the Frankfurter method often require additional treatment with applications of polyurethane in ethylacetate.[2] This is necessary because the rope is often extremely delicate after processing. Rope treated with the Frankfurter method retains its pretreatment color and the individual fibers, yarns, and strands that comprise the rope are well preserved. Like other successful treatments for severely deteriorated waterlogged rope, this process is generally not reversible due to two factors. First, the application of polyurethane in ethylacetate is generally not reversible. More important, however, is the fact that most treated rope samples are very friable and desiccated after treatment, making additional treatment difficult.

Treating Waterlogged Rope in a Nonpolar Suspension Medium

Experiments conducted by the National Museum Conservation Laboratories have also demonstrated that treating waterlogged rope with PEG in a volatile, nonpolar solution such as ether or kerosene enables the individual fibers to "float" during treatment, facilitating thorough impregnation of the PEG within the rope's matrix. The use of suspension mediums in PEG treatments results in rope specimens that lack the matted appearance of rope treated with PEG alone; the resulting rope, however, is extremely fragile and very susceptible to environmental changes.

Incorporating the Use of Nonpolar Suspension Mediums and Elements of the Frankfurter Method into "Traditional" Silicone Treatment Strategies

Experiments conducted with silicone oil treatments at the Archaeological Preservation Research Laboratory have demonstrated that treating waterlogged rope that has not been enclosed in some form of permeable material results in a specimen that tends to unravel slightly during treatment. We believed the polypropylene jacket used in the Frankfurter method would provide a permeable membrane that would facilitate chemical transfer and also serve to protect the physical integrity of the artifact during treatment. Furthermore, after observing the results of experiments conducted by the National Museum Conservation Laboratories on the use of suspension mediums in the treatment of waterlogged rope with PEG, we anticipated that the use of a suspension medium during the polymerization of waterlogged rope would alleviate the slightly matted appearance commonly observed after silicone oil treatments that do not involve the use of a nonpolar suspension medium.

Case Study: *La Belle* Rope

The following procedure is but one example of the use of preservation polymers in rope conservation. For researchers in the field of conservation, exploration with other silicone preservation polymers and cross-linkers is recommended in order to determine the resultant attributes of varying combinations of these invaluable materials.

The majority of rope recovered from *La Belle* was transported to the Texas A&M University Preservation Research Laboratory, where it was stored in freshwater until treatment. Three rope samples of similar length were taken from a single continuous coil. Two of these samples were to be treated by the proposed hybrid silicone treatment process (Si-1 and Si-2), while the third sample (WL) would be allowed to air-dry at room temperature for a 24-hour period.

The samples were rinsed in fresh running water for two days to ensure the removal of soluble salts. The samples were then placed on a sheet of glass for additional manual cleaning. During this process, the samples were positioned beneath a constant, gentle flow of tap water to keep the rope wet while debris was flushed from its surfaces. Like the majority of rope from the assemblage, the samples were partially covered with black and dark brown sulfide stains. These stains resulted from the fact that the cotton cloth in which the rope had been transported from the site to the laboratory had decayed en route. Most of these stains were removed by lightly rubbing the affected areas with a cotton swab. No attempts were made to remove deeply set stains by chemical means, as it was feared that additional chemical additives would interfere with the conservation process.

Each sample that was to undergo silicone treatment was placed between two sheets of perforated polyethylene film, which is scored with uniform holes that allow water, acetone, and silicone oil to freely diffuse (fig. 6.1). These sheets of polyethylene film were then heat-sealed, creating formfitting, ventilated bags in which the ropes would remain throughout the initial stages of treatment. Ziploc vegetable bags are an ideal source of perforated polyethylene film; they are readily available and easily sealed to form a pouch, using either a heat-sealing appliance or a small soldering iron and brown paper.

Each encased rope was placed into a beaker containing 500 ml of fresh acetone, and the beaker was set in a vacuum chamber. At room temperature, a vacuum of 3999.66 Pa (30 Torr) was applied to the samples to induce rapid displacement of the water with acetone. The samples initially bubbled rapidly as air and acetone were driven from the

Fig. 6.1. Waterlogged rope encased between sheets of Ziploc vegetable bag material prior to water/acetone displacement.

internal structures of the rope. After approximately 20 minutes, the rapid bubbling ceased, and smaller, more infrequent bubbles were observed escaping from the ropes. The samples were then removed from the water-laden acetone and placed into clean beakers containing 500 ml of fresh acetone. Each beaker was returned to the vacuum chamber and a vacuum of approximately 5332.88 Pa (40 Torr) was applied. Once the bubbling ceased, the ropes were removed from the vacuum chamber and allowed to sit at ambient pressure and room temperature.

The next phase of treatment was to exchange the acetone with an appropriate silicone/cross-linker solution. The polymer and cross-linker were specifically chosen to produce a desired texture and strength. To maintain flexibility in the treated rope samples, two hydroxyl-ended silicone oils were blended together in a 50:50 solution, by weight. The lighter of the two polymers was PR-10, which is a low viscosity hydroxyl-ended fluid.[3] Repeated experimentation indicates that lighter molecular weight silicone oils such as PR-10 tend to penetrate easily into organic materials such as rope; once polymerized, however, they tend to produce a rigid artifact.

PR-12 is a slightly more viscous hydroxyl-ended fluid with a larger molecular weight than PR-10. Because of the porosity of the waterlogged rope, larger molecular weight polymers such as PR-12 are expected to easily permeate the matrix of the rope samples. Due to its increased viscosity, PR-12 acts as a consolidant by keeping loose strands together; furthermore, rope that has been treated with PR-12 tends to be more flexible after treatment than rope treated with smaller molecular weight polymers. A blend of these two silicone oils was used for this experiment to ensure that the finished product maintained a degree of flexibility as well as internal rigidity and physical strength. CR-20 3% by weight was added to the PR-10/PR-12 silicone oil solution. CR-20 is a highly efficient cross-linker that has been shown to work well with silane polymers.

After placing the dehydrated ropes in clean beakers, a sufficient amount of the silicone oil/cross-linker solution was added to each beaker in order to immerse the samples in solution. Aluminum mesh was securely fixed over the packaged ropes to prevent them from floating to the surface of this viscous mixture. A vacuum of 5332.88 Pa (40 Torr) was applied to the samples in solution for 20 minutes to ensure that the acetone present in the rope fibers would vaporize rapidly, facilitating a thorough penetration of silicone oil solution throughout the artifacts. During the initial stages of vacuum treatment, large bubbles were observed escaping from the ropes. After 30 minutes, this rapid bubbling diminished and sporadic, small bubbles rose from the artifacts.

The packaged ropes were then taken out of the vacuum chamber and allowed to sit in solution at ambient pressure and room temperature. After 2 hours the samples were removed from the silicone oil/cross-linker solution and from their perforated polyethylene bags. The samples were placed on an aluminum screen to allow drainage of excess free-flowing silicone oil solution. After 1 hour the surfaces of the ropes appeared to be reasonably dry, and the artifacts were placed in beakers containing 500 ml of fresh CR-20 cross-linker. Immersion in CR-20 after bulking the samples with a silicone oil/cross-linker solution is helpful in removing additional silicone oil solution from the external surfaces of the rope. During the immersion process, the rope surfaces were wiped with a soft brush to facilitate removal of excess silicone oil solution. After five minutes of immersion and surface preparation, polymerization was initiated by exposing the rope sample to a tin-based catalyst.

The samples were placed into loose, perforated polyethylene bags, and the bags were heat-sealed shut. Each package was suspended with two wooden clothespins from wooden dowels. These dowels rested on the top edges of a small vat containing kerosene with CT-32, 3% by weight (fig. 6.2). The open structure of the mesh bag evenly exposed the surfaces of the rope to the kerosene/catalyst solution. With the samples suspended in the solution, the vat was placed into a vacuum chamber, and a vacuum of 5332.88 Pa (40 Torr) was applied. After 20 minutes under vacuum, the valves of the chamber were locked, and the rope was left suspended in the solution overnight. The following morning, the vacuum chamber was returned to ambient pressure, and the samples were removed from the kerosene/catalyst solution.

The rope was removed from the perforated polyethylene bags and placed on several paper towels, which absorbed the kerosene/cross-linker and silicone oil solutions from the artifacts. Immediately following removal from the kerosene/cross-linker solution, the surfaces of the cordage were covered with a thin, slippery coating of silicone. After the rope was exposed to fresh air for a few minutes, droplets of fully cured polymer were observed on one end of the samples. These were easily removed using a soft, lint-free cloth. After the rope was allowed to air-dry in a vented fume hood for 24 hours, the surfaces of the samples appeared dry and very natural in texture. To determine the degree of deterioration caused by waterlogging, as well as to compare the results of the silicone-treated rope against an untreated specimen, an additional sample of rope from the same coil was weighed, measured, and allowed to air-dry at room temperature in a vented fume hood for 24 hours.

Results and Discussion

The silicone oil–treated rope samples felt slightly stiff, but were stable and aesthetically pleasing compared to the sample that was simply allowed to air-dry. The individual fibers, strands, and ply of the silicone-treated rope samples were easily distinguishable after treatment and did not become matted or compressed. The high degree of visible detail in the silicone-treated samples was surprising because these features were indistinguishable in a waterlogged state. And while it may be impossible to determine the original color of the waterlogged rope, the post-treatment coloration of the silicone-treated samples was acceptable, ranging from a pale to medium brown. (See tables 6.1, 6.2, 6.3, and 6.4.) Figure 6.3 is a photograph of sample Si-1. Prior to treatment, this sample was loosely twisted. Following treatment, no discernible changes were observed in the twist or physical dimensions of either silicone oil–treated rope sample.

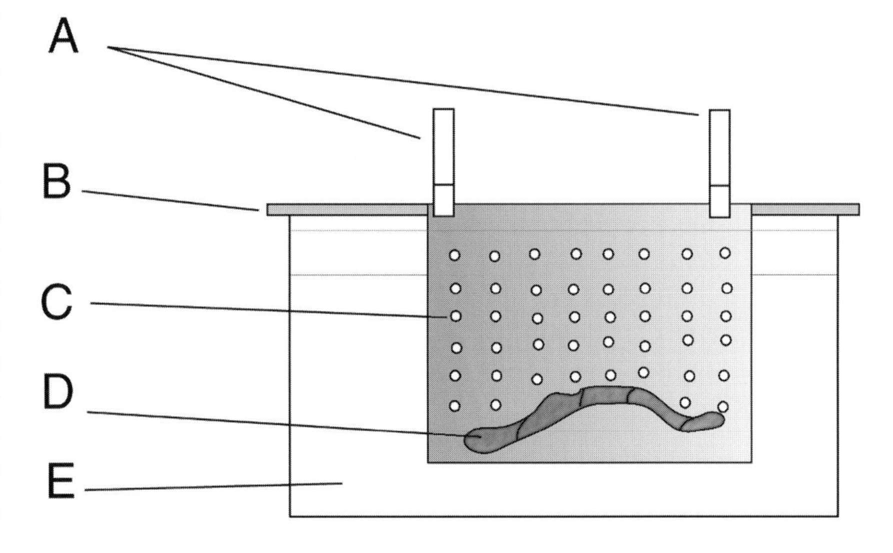

Fig. 6.2. Rope sample suspended in kerosene/CT-32 solution: *(A)* clothespins; *(B)* dowel spanning the length of the vat; *(C)* perforated polyethylene bag; *(D)* silicone-impregnated rope suspended in mesh; *(E)* vat containing kerosene/CT-32 solution.

Posttreatment Strength

After one week of air-drying, the silicone oil–treated rope and the comparably sized sample of untreated rope (WL) were taken to the Texas Engineering Experiment Station, Testing Machinery and Repair Laboratory at Texas A&M University for tensile strength testing. We believed these tests would provide insight into the strength characteristics of polymer-treated rope. Tensile strength testing was conducted with a 20 kip (1 kip = 1000 lbs tensile strength) MTS servohydraulic frame, which measures the maximum load-breaking point of materials. To more accurately measure the maximum load-breaking point of low-potential tensile strength materials such as the fragile, treated *La Belle* rope fibers, a 2 kip load cell was mounted into the jaws of the 20 kip machine.

Data control and acquisition were recorded using Gardner Systems software. Time, distance, and pounds force were measured for each sample. In each test, tensile strength testing continued until the sample failed. Rope 1 (Si-1), treated with silicone oils, was mounted in the load frame using wedge grips. This sample slipped once during testing. To prevent slippage with the other samples, the second silicone-treated rope sample (Si-2) and the untreated sample (WL) were mounted into the load frame using wedge grips only after being outfitted with epoxy potted ends. This is a more complex mounting process that requires that the ends of the rope be cemented into a cone-shaped epoxy base prior to mounting in the frame. The use of these potted ends eliminated slippage and resulted in more reliable data.

Tensile strength testing has demonstrated that rope preserved in silicone oil is considerably stronger than rope that has been allowed to air-dry. When tested, the rope section WL failed at 2.6 lbs tension, and the sections of rope treated with silicone oils, Si-1 and Si-2, failed at 36.5 lbs and 27.7 lbs, respectively. Table 6.5 lists the data acquired from this tension test. As a result of waterlogging, which deteriorated and weakened individual fibers of the rope, the strands that make up the rope failed at different times in each of the samples. The sawtooth graphs in figures 6.4, 6.5, and 6.6 represent the tension and breaking points of the fibers, strands, and plies of the rope samples. Although posttreatment strength may not be an important factor in the deci-

Table 6.1 Measurements and Percentage Changes for Air-Dried Rope Sample WL

Time	*Weight (in g)*	*% Change*	*Width (in cm)*	*% Change*	*Length (in cm)*	*% Change*
Before air-drying	12.2	—	1.516	—	9.788	—
1 hour air-drying	3.10	−74.590%	1.480	−2.374%	8.761	−10.492%
3 hours air-drying	2.82	−76.885%	1.472	−2.902%	8.751	−10.594%
5 hours air-drying	2.70	−77.868%	1.472	−2.902%	8.642	−11.708%
24 hours air-drying	2.70	−77.868%	1.472	−2.902%	8.531	−12.842%

Table 6.2 Texture, Integral Strength, and Color of Air-Dried Rope Sample WL

	Pretreatment	*Posttreatment*
Texture	Soft-mushy	Brittle
Integral strength	Fragile	Friable
Color	10YR-2/2 very dark brown → 10YR-2.1 black	5YR-6/2 → 5YR-5\2 pinkish gray

Table 6.3 Measurements and Percentage Changes for Silicone-Treated Rope Samples

	Measurement	*Mean*	*% Change*
Length of rope (in cm)			
pretreatment	14.308	14.228	−1.118
posttreatment	14.148		
Width Si-1 (in cm)			
pretreatment	0.9280	.922	−1.293
posttreatment	0.9160		
Width Si-2 (in cm)			
pretreatment	0.9380	.932	−1.386
posttreatment	0.9250		
Weight of rope (in g)			
pretreatment	16.597	11.499	−61.438
posttreatment	6.400		

Table 6.4 Color and Flexibility of Silicone-Treated Rope Samples

	Pretreatment	*Posttreatment*
Color	10YR-2/2 very dark brown → 10YR-2.1 black	10YR-6/3 pale brown → 10YR→ 4/3brown
Flexibility	Limp, almost formless, individual strands indistinguishable	Dry, individual strands visible, slightly stiff

Note: Color determined using Munsell Soil Color Charts, Macbeth Division, Kollmorgen Corporation, Baltimore, Md., 1975 ed., p. 26.

sion to conserve rope by a particular method, it is beneficial to know that silicone oil–treated cordage is more internally stable and stronger than rope not treated with silicone oil.

Effectiveness of Incorporating Nonpolar Suspension Mediums and Elements of the Frankfurter Method into "Traditional" Silicone Treatment Strategies

The Danish process of using perforated polypropylene film to make a formfitting jacket within which archaeological rope is treated—an approach most commonly utilized in the Frankfurter method—worked well when incorporated into our silicone treatment. The polypropylene jacket provided a permeable membrane that facilitated chemical transfer and also helped protect the physical integrity of the artifacts during treatment. Rope treated without being enclosed in some form of permeable material, such as a perforated polyethylene bag, tends to unravel slightly during treatment.

Silicone oil–treated rope treated without a nonpolar suspension medium such as kerosene usually results in an artifact with a

Fig. 6.3. Silicone oil–treated rope Si-1.

Table 6.5 Tensile Strength Data for Rope Samples

File	*Sample*	*Peak Load (in lbs)*	*Comments*
Rope 1	Si-1	36.5	Wedge grip mounted
Rope 2	Si-2	27.7	Epoxy potted ends
Rope 3	WL (waterlogged)	2.6	Epoxy potted ends

slightly matted appearance. Immersion in a nonpolar solution enables the fibers to "float" and facilitates the polymerization of individual fibers. The use of kerosene with a tin-based catalyst works well as a medium for polymerization, but the kerosene/catalyst polymerization medium is not ideal for routine laboratory use due to the flammable nature of kerosene. An additional disadvantage in using a kerosene/catalyst medium for polymerization is that it takes several days for the faint odor of kerosene to be eliminated from the artifact. Other volatile solvents may work well in place of kerosene, but these experiments have not yet been conducted.

Experimentation has indicated that CR-20 cross-linker/CT-32 catalyst, 3% by weight, solution is an effective and safe substitute for the kerosene/catalyst mixture. After treated samples were removed from the CR-20/CT-32 catalyst solution, the surfaces of the rope were not slippery, suggesting that more complete catalysis occurred while the sample was in solution. Residual odors associated with this catalyzation medium dissipated in a matter of minutes once the sample was exposed to fresh air.

After organic materials have been impregnated with silicone oil/cross-linker solutions, immersing them in CR-20 cross-linker and surface-wiping with a cotton swab or a lint-free cloth is an effective way to remove excess polymer from the surfaces of an artifact. In some cases, excess silicone oils have been removed from the surface of rope by immersion in CR-20 under a slight vacuum. The process appears to eliminate a great deal of silicone oil solution from within the deep crevices and voids on the rope's surface.

Obvious benefits of using silicone oils for conserving waterlogged rope include the short treatment duration and the minimal amount of laboratory equipment required for the process; PEG/freeze-drying methods of rope preservation require substantially more time and labor. The silicone oil–treated examples were conserved in less than 24

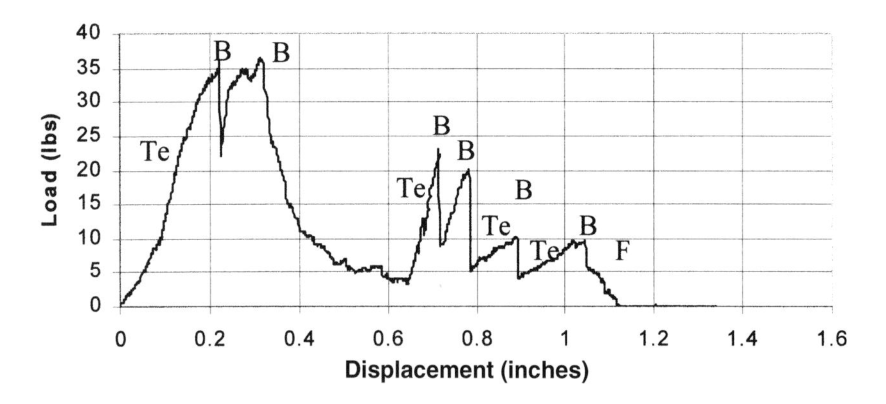

Fig. 6.4. Rope Si-1 load/displacement data: *Te* increasing tension; *B* break point of strand; *F* failure of strand.

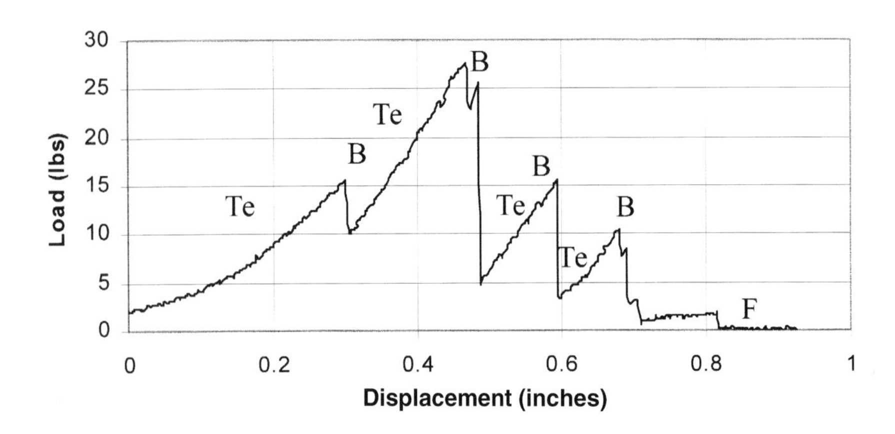

Fig. 6.5. Rope Si-2 load/displacement data: *Te* increasing tension; *B* break point of strand; *F* failure of strand.

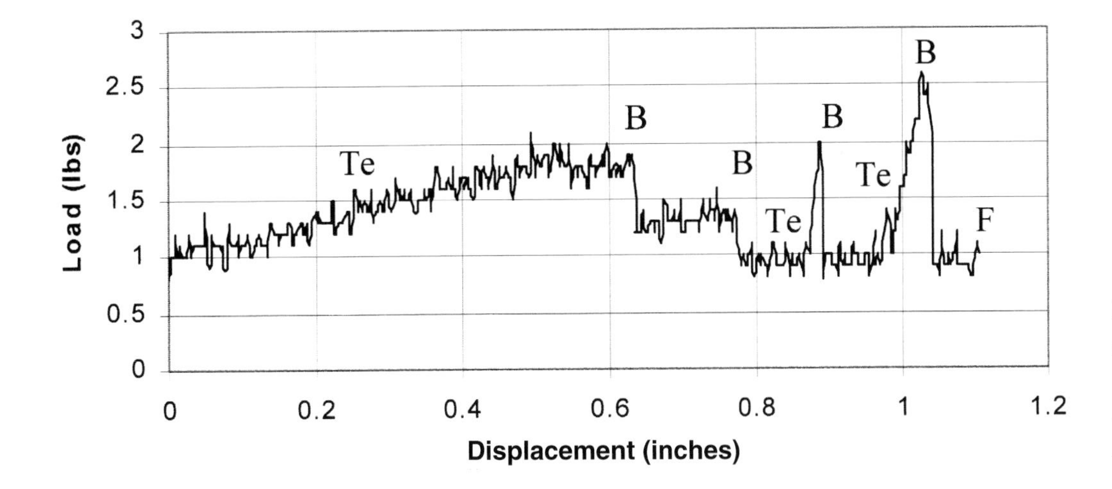

Fig. 6.6. Rope WL load/displacement data: *Te* increasing tension; *B* break point of strand; *F* failure of strand.

hours. In addition, artifacts treated with silicone oils do not require special curation and the dimensional attributes of the artifact seem to be accurately preserved. As Vera de la Cruz Baltizar has observed, silicone oil–treated samples appear to be dimensionally stable with good coloration.[4] Repeated testing of the hybrid silicone oil treatment described above at the Archaeological Preservation Research Laboratory consistently yields waterlogged rope specimens that are aesthetically pleasing and dimensionally stable.

Accelerated aging tests and data supplied by Dow Corning Corporation continue to be encouraging regarding the long-term stability of silicone oil–treated artifacts. Eight silicone oil–treated samples were subjected to an extended test in an accelerated weathering machine (fig. 6.7). The samples were exposed to four months of continuous, alternating cycles of 6 hours at high humidity (95%) and high temperature (45°C) with a UV 340 light, and 6 hours at a lower humidity (60%) and temperature (20°C) with no light exposure. The tested sample data (including overall dimensions, color, and surface integrity) were comparable to data for silicone oil–treated specimens that had not undergone accelerated weathering.

Silicone oil treatments, like treatments that require applications of polyurethane in ethylacetate, are not reversible; this is not to say, however, that rope treated with silicone oil cannot be re-treated. In the past, several fragile leather and canvas artifacts preserved with lower centistoke silicone oils have been

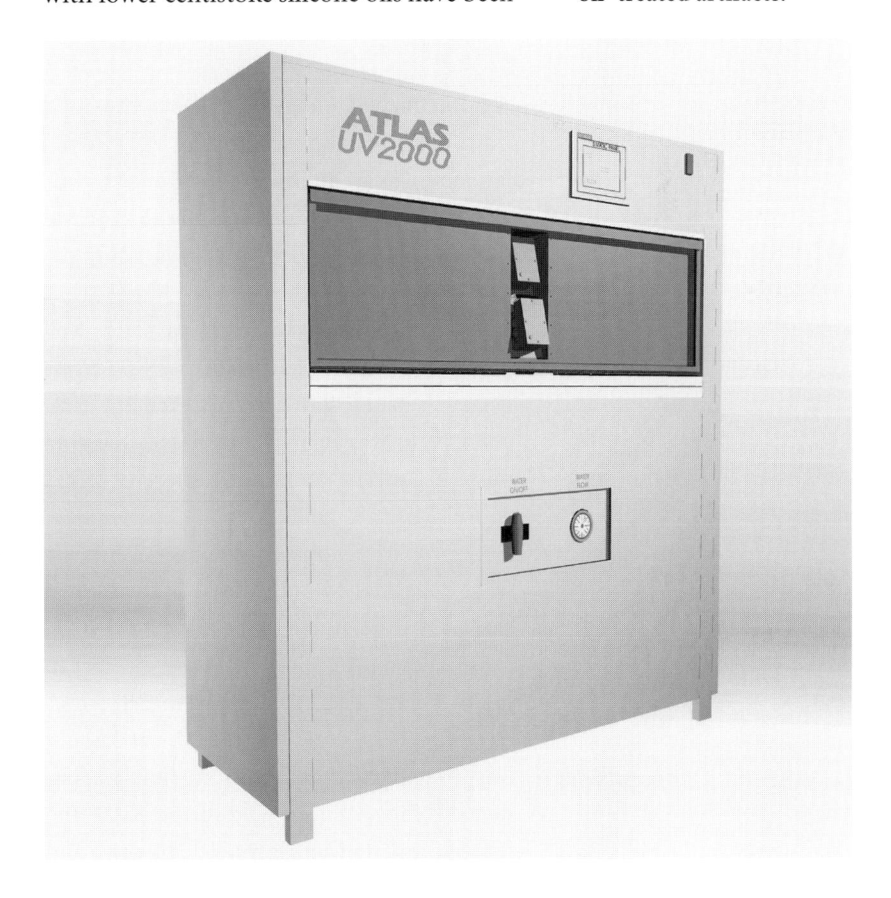

Fig. 6.7. Accelerated weathering machinery used for testing silicone oil–treated artifacts.

re-treated, using more viscous polymers to add additional strength and stability. More important, waterlogged rope appears to respond well to treatment using silicone oils. Although silicone oil processes will not address all archaeological conservation issues, the field of archaeological conservation can benefit from ongoing research into silicone oil treatment techniques.

Based on experimentation to date, I concur with Baltazar's observations that silicone oil preservation is a promising technique for the consolidation and preservation of many waterlogged materials.[5] The added strength and elasticity characteristics associated with the silicone oil process used for this experiment may have important implications for the structural well-being of some artifacts. Continuing research at the Archaeological Preservation Research Laboratory is focused upon these issues.

Case Study: Preservation of Waterlogged Canvas from Port Royal

Among the wide range of artifacts recovered from underwater excavations at Port Royal, Jamaica, was an interesting section of a gudgeon from a vessel. Because the Port Royal site is covered by a deep silt layer, many organic artifacts have been recovered that would not have survived, or at least be as well preserved, at a land site. Artifact PR90 2074-17 is such an object. When recovered, the gudgeon plate was heavily concreted, looking more like a long rock than a piece of ship's equipment.

During initial phases of conservation, we were delighted to note that a long strip of canvas had been used as a backing behind the gudgeon plate, along with a thick layer of pitch. After the outer layer of concreted material had been removed from the artifact, it was a relatively simple task to slowly separate the canvas backing from the back surface of the plate. Once removed, the cloth and metal plate could be treated using appropriate conservation strategies with the idea that, once completed, the gudgeon assembly could be documented and reconstructed on paper.

Microscopic analysis of the canvas indicated that the backing had been made from cotton canvas material that had been crudely cut to follow the shape of the gudgeon plate. Two squared holes were present in the section that was to be treated using silicone oils. These holes lined up with the remains of two large bolts that would have been used to fasten the gudgeon plate to the hull of the vessel. The warp of the fabric was counted at 20 strands per inch while the weft averaged 16 strands per inch. Variances in warp and weft counts, as well as the generally uneven shape of the strands, suggest that this was probably a common, though sturdy, material.

Once removed from the back of the gudgeon, the canvas was placed into a vat of fresh tap water, and the surfaces of the cloth were lightly cleaned, using fingertips to remove loose debris and concretion. The canvas was initially placed into a vat of acetone, which successfully removed traces of pitch embedded in the weave of the fabric. After several rinses of freshwater, which were used to remove salts and any remaining acetone from the fabric, the canvas was placed into a 5% solution of hydrochloric acid in water to aid in the removal of rust stains and minute specks of concretion. After treatment, the cloth was immersed in several freshwater baths to remove any remaining acid. The canvas was then briefly placed into a 5% solution of hydrogen peroxide to remove heavy sulfide stains. The fabric was again rinsed in freshwater.

The goal of the entire cleaning process was to remove sufficient pitch and staining to enable the polymer to penetrate the cloth but not enough to disturb their function as diagnostic attributes of the canvas. The cleaning process complete, the fabric was placed into an acetone bath to dehydrate the material in preparation for silicone bulking. To aid in the removal of any remaining water, the container holding the fabric in acetone was placed into a vacuum chamber and a vacuum of 28 Torr was applied until all bubbling ceased. At this point, all remaining water had been removed from the artifact. The canvas was then quickly placed between two sheets of lint-free paper and blotted to remove some of the acetone in the fabric.

After blotting, the canvas was placed into a large beaker containing 500 mg of PR-12 silicone oil. A mesh screen was placed on top of the fabric in solution as a means of keeping the cloth suspended in solution throughout the bulking process. The beaker was then placed back into the vacuum chamber, and as before, a vacuum was applied to the fabric for 2 hours. Vigorous bubbling ceased after 20 minutes and no bubbles were noted after 1 hour of applied vacuum. The fabric was allowed to sit in solution and slowly returned to ambient room temperature, where it remained for 48 hours.

Before the fabric was removed from the silicone solution, a warming oven was preheated to 48°C. As in many of our other experiments, a containment chamber was created by placing an inverted polypropylene pail with a tight-fitting lid into the warming oven. In the center of the lid (acting as the base of the chamber), we placed a flat dish containing 57 grams of CT-32 catalyst. A large mesh screen covering the dish served as a platform on which the fabric could be placed for exposure to warm catalyst fumes (fig. 6.8). After much of the free-flowing silicone was removed by suspending the canvas over a stainless steel dish and allowing excess silicone to drip from the material, the fabric was placed between two sheets of newspaper and lightly pressed to remove additional oils. The fabric was then placed onto the screen in the containment chamber, where it was exposed to catalyst fumes for 24 hours. After 24 hours, the catalyst had become hard and crusty as a result of silicone oils dripping into the dish, forming polymerized material. A new tray containing 57 grams of catalyst was placed into the containment chamber, and the process was continued for an additional 24 hours.

After two days of polymerization, the fabric felt slightly damp, but had retained the look and feel of cotton canvas material. The warming oven was turned off, and the fabric was left in the containment chamber at room temperature for an additional 24 hours. After a total of three days, the fabric was removed and allowed to air-dry undisturbed on a sheet of white paper.

Observations

The processes used for removing embedded concretion and stains from the cotton canvas are conventional conservation processes that are widely used in the treatment of stable fabrics. Care was taken to thoroughly rinse the artifact prior to bulking with silicone oils; the effects of interaction between these materials are not known. The porous nature and loose weave of the fabric suggested there would be no difficulty in the dehydration and subsequent bulking of the material with silicone oils; this meant that using the longer process of applying a vacuum to the fabric

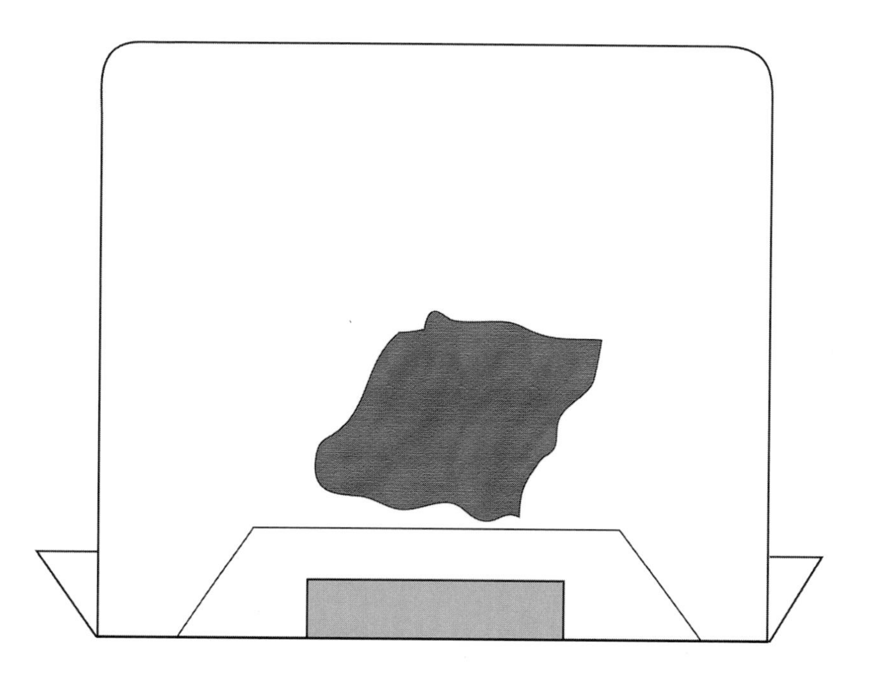

Fig. 6.8. Containment chamber configuration illustrating the canvas material sitting on an aluminum screen above a tray containing CT-32 catalyst.

Fig. 6.9. Posttreatment view of the canvas after catalyzation.

in a freezer-mounted vacuum chamber to drive off acetone was not necessary.

Pre-and posttreatment traces of the outline of the canvas material, as well as measurements taken from the edge of the material to the edges of the nail holes, indicated that no determinable shrinkage occurred in the fabric. More extensive treatment in acetone baths prior to silicone bulking would have removed more of the staining and pitch from the fabric. These, however, are diagnostic attributes of the cloth, indicating something about the mounting of the gudgeon strap to the vessel. Accordingly, dehydration time was closely monitored to ensure that these stains were not removed.

After treatment, the fabric was both flexible and stable. Indeed, five years after treatment, flexibility and physical integrity of the canvas remain unchanged (fig. 6.9). Placing the treated canvas on sheets of white paper was the simplest means of determining the degree of polymerization that had occurred. After it sat on paper for one week, no oil spots were noted. If polymerization were incomplete, the points of contact between the canvas and paper would be oil stained. Since no stains were noted, this was a good visual indicator that polymerization was complete.

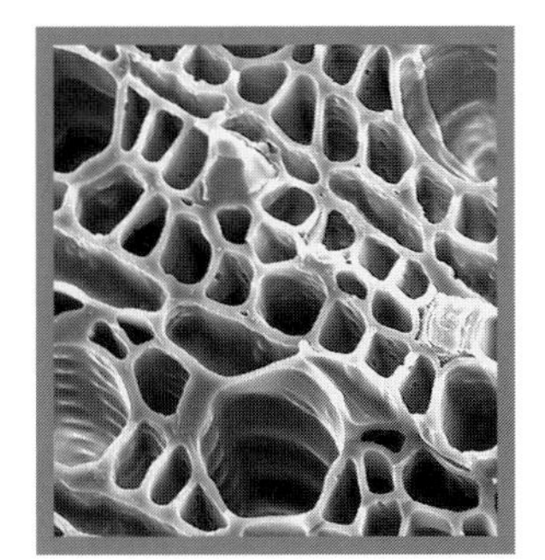

CHAPTER 7

Glass Conservation

In ancient times, artisans combined secret ingredients, which they heated to high temperatures to form a wide range of colors of glass. They used ash and varieties of plants to make flux, which helped in melding their complex chemical compositions. The chemical processes in a marine environment that act to deteriorate glass are as complex and varied as the processes of making it. In this environment, organic components of ancient glass are easily leached out of its matrix through a series of complex reactions, rendering the glass unstable.

Despite the fact that glass has been made for more than 3500 years, scientists know little about the development of the glassmaking process and the myriad combinations of materials used by artisans to create glass throughout history. Archaeologists have recovered an incredible variety of glass from terrestrial and underwater excavations. From analysis of these artifacts, we know that a wide variety of raw materials can be combined to produce raw glass.

Although there is significant debate about the fundamental nature of glass, Robert Brill of the Corning Glass Museum considers it to be a "state of matter that combines the rigidity of a crystal with the largely random molecular structure of a liquid."[1] He has observed that hundreds of thousands of forms of glass have been made; the term "glass" therefore is a generic term used to loosely describe an endless series of chemical variations. This variety poses a considerable challenge for archaeological conservators.

The basic component of glass is silica (SiO_2), which, unlike quartz, is a randomly arranged network of crystalline structures. The presence of additional elements such as oxides (R_2O) or potash (K_2O) act as fluxes, which by their nature affect some silicon-oxygen bonds. Unattached oxygen atoms are negatively charged and, as Cronyn has observed, loosely hold monovalent cations within the spaces of the network.[2] Because these bonds are weak, the cations migrate within and out of the network in close association with water. The proportion of silica, flux, and other stabilizing agents within glass is infinitely variable. These proportions affect the character and quality of glass. Additionally, the combination of these elements will determine the melting point of the glass. The coloration of glass depends on the state of oxidation-reduction of the artifact as well as mixtures of ions and the presence of additives, which can accentuate or diminish color.

The basic composition of nearly all ancient glass consists of a soda-lime-silica mixture with substantial impurities added to the mixture from raw materials. In medieval Europe, the source of silica was sand and, in some cases, crushed stones. Limestone and burned shells were a common source for the lime component of glass. Mesopotamian tablets also specify that "ground red shells from the sea" were a good source for lime.[3] As the

Roman Empire expanded, so did the spread of technologies related to glassmaking. This influenced English glassmaking and the technologies used during the colonial period of Jamaica.

Numerous factors complicate attempts to understand the processes of devitrification that affect archaeological glass. First, the interaction of water with glass, especially seawater, causes chemical instability. Association with other materials within an archaeological site can complicate the chemistry of glass, since oxides and minerals may interact positively or negatively with the degradation processes of the glass. In addition, the sediments in which the glass is buried profoundly affect the degradation process. These and other factors account for the wide range of colors and degrees of degradation observed within the Port Royal glass assemblage, consisting of hundreds of bottles, known as onion bottles; sections of window pane, often with lead caming attached; and fragments of small ointment bottles.

Glass is usually the most stable of archaeological materials, but it can undergo some complex disintegration—especially seventeenth-century glass. Ideally, glass should consist of 70–74% silica, 16–22% alkali or soda ash (sodium carbonate) or potash (potassium carbonate, usually derived from wood ash), and 5–10% flux lime (calcium oxide). Soda-lime glass is the most common glass throughout the history of glassmaking, and the modern equivalent is approximately 65–75% silica (SiO^2), 12–18% soda (Na_2CO_3), and 5–12% lime added as a stabilizer. Soda glass is characteristic of southern Europe, where it is made from crushed white pebbles and soda ash derived from burnt marine vegetation. Soda glass, often used for the manufacture of cheap glass, is twice as soluble in water as potash glass.

Potash glass is more characteristic of inland Europe, where it is made from local sands and potash derived from wood ash and burnt inland vegetation. A little salt and minute amounts of manganese are added to make the glass clear, but potash glass is less clear than soda glass. Most early glass is green because of iron impurities in the materials. The alkali lowers the melting point of the sand and the flux facilitates the mixture of the components. As long as the original glass mixture is kept in balance, the resulting glass will be stable. Problems arise when an excess of alkali and a deficiency in lime (calcium oxide is used as a stabilizer) are used in the mixture, for the glass will be especially susceptible to attack by moisture. A relative humidity (RH) higher than 40% can be dangerous. Old glass with 20–30% sodium or potassium may begin to weep and break down. This is often called glass disease.

In all glass, the sodium and potassium oxides are hygroscopic; therefore, the surface of the glass absorbs moisture from the air. The absorbed moisture and exposure to carbon dioxide causes the NaO_2 or NaOH and the KO_2 or KOH to convert to sodium or potassium carbonate. Both $NaCO_2$ and KCO_2 are extremely hygroscopic. At an RH of 40% and above (and in some cases as low as 20%), drops of moisture appear on the glass surface. In water, especially saltwater, the Na and K carbonates in unstable glass can leach out, leaving only a fragile, porous hydrated silica (SiO_2) network. This causes the glass to craze, crack, flake, and pit, giving it a frosty appearance. In some cases, an actual separation of layers of glass from the body occurs. Fortunately, glass from the nineteenth century or later seldom manifests such effects. Problems are rarely encountered on glass found at sites dating from this period. Pearson discusses glass deterioration and reviews the various glass conservation procedures.[4]

Currently, the decomposition of glass is imperfectly understood, but most glass technologists agree that glass decomposition is due to preferential leaching and diffusion of alkali ions (Na and K) across a hydrated porous silica network. Sodium ions are removed and replaced by hydrogen ions, which diffuse into the glass to preserve the electrical balance. The silicates are converted into a hydrated silica network through which sodium ions can diffuse out. Glass retrieved from an acid environment often appears laminated, with an iridescent film formed by the leached silica layers. The alkali that leaches out is neutralized by the acid and fewer hydroxyl ions are available to react with the silica. This causes the silica layer to thicken and become gelatinized as the alkali leaches out. Glass excavated from an alkaline environment is less likely to have laminated layers because an abundance of hydroxyl ions are available to react with the silica network. Normally, a protective layer does not form on glass exposed to alkaline solutions. The dissolution of the glass proceeds at a constant rate. The alkali ions are always extracted in excess of the silica, leaving an alkali-deficient layer that continually thickens as the deterioration moves deeper into the glass.

There are considerable differences of opinion regarding the preservation of unstable glass. Some conservators believe the only treatment should be to keep the glass at low RH so that it cannot react with any moisture. While a RH environment of 40% is ideal for archaeological class, the range can be extended to 55% if the glass is stable. The weeping or sweaty condition is sometimes made worse by the application of a surface lacquer or sealant, which traps humidity in close association with the affected glass. No resin sealants are impervious to water vapor and the disintegration continues under the sealant until the glass falls apart. Other glass conservators try to remove the alkalinity from the glass to halt the deterioration.

When it comes to eighteenth- and nineteenth-century sites, we are on much more secure ground. Most, if not all, of the glass found from this period was produced from a stable glass formulation, with few problems arising other than normal devitrification. Since the glass is impervious to salt contamination, no conservation treatment other than simple rinsing and removal of incidental stains (especially lead sulfide staining on any lead crystal) or calcareous deposits is envisioned. The main problems will be related to gluing pieces together and possibly some restorations. All the problems likely to be encountered are discussed thoroughly in *Conservation of Glass* by Roy Newton and Sandra Davison.[5]

Submersion in seawater will cause glass to become unstable. If left untreated, the artifact will deteriorate. A technique described by Plenderleith and Werner is representative of a range of conservation strategies that are effective for stabilizing archaeological glass.[6]

1. Wash the artifact thoroughly in running tap water.
2. Soak it in distilled water.
3. Immerse the artifact in two baths of alcohol to dry quickly. This treatment will retard the disintegration and also improve the appearance of the glass. It does not, however, always stop the breakdown of the glass.
4. If applicable, apply an organic lacquer—PVA, Acryloid B-72—to impede the disintegration.
5. Store the artifact in a dry environment with the relative humidity no higher than 40%; others say 20–30% is ideal. The Corning Glass Museum keeps incipient crizzled glass stored at 45–55% RH. RH 42% is the critical point at which KCO_3 becomes moist.

The above treatment does not attempt to remove any of the glass corrosion products, which often result in layers of opaque glass and which can be removed with various acid treatments. The decision to remove surface corrosion products that often mask the color of the glass must be made on a case-by-case basis. Removal of corrosion products can significantly reduce the thickness of the walls and sometimes weaken the glass object. Indiscriminate removal of surface corrosion products can weaken, blur, or alter surface details. The corrosion layers of a glass object can be deemed a part of the history of the object and thus a diagnostic attribute. They should not be removed without good reason.

Devitrification

Devitrification is a natural process that occurs in siliceous material. In flint and obsidian, this process provides the basis for obsidian hydration dating. The surface of any glass from any time period usually becomes hydrated, especially soda glass. Devitrification occurs when the surface of the glass becomes partly crystalline as it absorbs moisture from the atmosphere—and from being submerged in water. As the glass becomes crystalline, the surface becomes crazed and flakes from the body. Devitrified glass has a frosty or cloudy, iridescent appearance. Pane glass is especially susceptible to this process. To prevent further devitrification and to consolidate the crazed surface, a coating of PVT or Acryloid B-72 is applied. Any of these surface adhesives will smooth out the irregularities in the pitted, crazed surface of the glass, making it appear more transparent by filling in the small cracks and forming optical bridges. Merely wetting glass will cause it to appear clearer for the same reason.

Removal of Sulfide Stains from Lead Crystal

Leaded glass, which includes a wide variety of stemware and forms of lead crystal, can become badly stained by lead sulfide. Glass that is normally clear will emerge from marine and anaerobic sites with a dense black film on its surfaces. A 10–15% solution of hydrogen peroxide is used, as with ceramics, to remove these sulfide stains. Besides stain removal, strengthening of glass artifacts with a consolidating resin is often required. Fragments can be reassembled with a good glue or, if necessary, an epoxy such as Araldite.

Consolidating Waterlogged Glass Using Passivation Polymers

Another useful method for halting devitrification is treating waterlogged glass with silicone oil, which, when polymerized, forms a complex layer and bond with the remaining matrix of the glass. As noted in the Plenderleith and Werner process, the first steps must include extensive washing, first in tap water and then distilled water, to remove soluble salts and debris. After rinsing, the following steps have proved effective in consolidating small glass artifacts with silicone oils.

First, it is essential to drive off water from the artifact. This can be accomplished by using two baths of alcohol and allowing the glass to remain immersed in each bath for 24 hours. For processes utilizing silicone oils, additional dehydration using fresh industrial-grade acetone is beneficial. To prepare glass for treatment with a silicone oil solution, the most effective means of water/acetone exchange is to use vacuum to assist in driving off water in the waterlogged glass. Acetone is more volatile than alcohol, which assists in removing water and alcohol from the glass object.

Used in conjunction with a vacuum chamber, the process of acetone/silicone oil displacement is much more efficient than alcohol/silicone oil displacement. When a vacuum is applied to glass in treatment, the boiling point of acetone within the matrix of the glass is lowered, causing remaining acetone to be rapidly driven from the glass. This creates a void, enabling silicone oils to replace water and acetone that have been removed from the artifact.

An Effective Silicone Oil Treatment Strategy

The following is a simple and effective process for treating small glass artifacts. It is essential that most of this work be conducted in a well-ventilated fume hood or work area. Rubber gloves should be worn to prevent long-term chemical contact with bare skin. Older vacuum pumps, which have all-metal components, are more desirable than newer pumps when working with acetone; vapors escaping into these pumps will not damage any internal parts. If a modern pump is used, a Dewar flask, gas traps, and dry ice must be used to prevent acetone fumes from damaging the plastic components in the pump.

1. After immersing the artifact in a container of fresh acetone, place the container in a vacuum chamber. Slowly increase the vacuum in the chamber. Initially, there will be a profusion of large bubbles (fig. 7.1). Over a short period of time, rapid bubbling will subside, though smaller bubbles will continue to be visible. At this point, acetone and remaining water are being driven from the artifact. The valves of the vacuum chamber should then be locked off and the artifact allowed to sit in the solution for at least 1 hour.

2. While water/acetone exchange is taking place, a suitable consolidant should be prepared. For heavily crizzled glass, PR-14 has proved to be an excellent polymer. Enough PR-14 should be poured into a clean, dry container to ensure that the artifact will be completely submerged during treatment. For small artifacts, plastic-coated paper cups are useful and convenient. To this volume of PR-14, add a 3% addition of CR-20 (by weight). This solution needs to be mixed thoroughly, using a nonporous stir rod. Rapid mixing will result in bubbles in the solution. These can be removed, if desired, by placing the polymer solution into a vacuum chamber and applying a slight vacuum.

3. Once the polymer solution is ready, carefully and quickly remove the artifact from the acetone and immerse it in the PR-14/CR-20 solution. A small piece of aluminum screen over the glass will prevent it from moving during treatment. Place the glass in solution into a vacuum chamber and slowly apply a vacuum. Almost immediately, small bubbles will begin emerging from the artifact. As acetone is driven from the artifact, silicone oil will displace it. A vacuum of 28 Torr (3733.016 Pa) is sufficient to assist penetration of the consolidant into the matrix of the glass. After allowing the glass in solution to sit at 28 Torr (3733.016 Pa) vacuum for 30 minutes, slowly return the artifact in solution to ambient pressure. Allow the artifact to sit in solution for an additional hour before handling (fig. 7.2).

4. For the last phase of treatment, two small beakers are needed. The first beaker should be filled with sufficient CT-32 catalyst to immerse the artifact. An equal volume of CR-20 cross-linker should be poured into the second beaker. Once the cross-linker- and catalyst-filled beakers have been prepared, the artifact should be carefully removed from the polymer/cross-linker solution and placed into the beaker containing the CR-20 cross-linker. Allow the artifact to sit in this solution for a few minutes. It should then be carefully removed and allowed to sit on lint-free cloth for a short period of time. It is difficult to give an exact time frame for allowing the artifact to drain, but usually two or three minutes is sufficient. After draining, the artifact should be placed in the CT-32 catalyst for a short period of time. Monitor this phase of treatment carefully. After a few seconds, a zone of chemical reactivity will be noticed around the artifact. Generally, the artifact is left in the catalyst for the same period of time as it was treated in the CR-20 solution. It should then be removed, very lightly surface-wiped with lint-free cloths, and the processes of immersion in CR-20 and CT-32 should be repeated, at least one more time. For most artifacts, two or three immersion cycles are sufficient to create a stable, consolidated end product (fig. 7.3).

5. After treatment, the CT-32 catalyst needs time to work. Some experimentation is required but, generally, only a very light surface-wiping of the artifact is required to remove free-flowing silicone oil solution from its surfaces. The glass should then be placed into a well-ventilated environment and allowed to air-cure for several hours before handling. It is possible to touch completed artifacts within a few minutes, but for a complete cure, the process may take as long as 12 hours. For larger artifacts, or in cases where additional contact with an active catalyst may be desirable, the finished artifact and a cloth with several drops of CT-32 can be placed into a glass jar with a tight-fitting lid. When the jar is tightly sealed, the artifact will react with the catalyst fumes, causing additional or more rapid polymerization.

Posttreatment Aesthetics

After treatment, the glass may require additional cleanup to remove surface deposits of silicone. In most cases, pooled silicone can be easily removed with a soft cloth. For larger surface pooling, wipe the artifact surface with a soft cloth containing a few drops of CR-20 cross-linker. This is usually sufficient to remove surface polymer and leave the artifact with a uniform, satin finish. If this doesn't remove pooled silicone from the artifact, use the edge of a soft wooden dowel or toothpick to gently mechanically clean the polymer from the artifact's surface.

Fig. 7.2. The glass bead in solution during the acetone/polymer displacement process.

Fig. 7.1. Rapid bubbling during the acetone dehydration process.

Analysis of Polymer-Treated Glass

Analysis of silicone-treated glass indicates that a complex, consolidating coating has been formed on the surface of treated glass. Chemical analysis of the glass also indicates that silicone polymers are present in all the test sites within the core of the sample. Figure 7.4 is a cross section of a piece of treated glass. Three distinct layers are visible in the cross view.

Reconstruction

Glass can be repaired and reconstructed with the same glues used for pottery. However, optically clear epoxy resins are generally used because they adhere to the smooth, nonporous glass more readily; they dry clearer and shrink less than the solvent resins; and they are, therefore, less noticeable. The bonds are also stronger. The epoxy resins, however, are usually irreversible. Hysol Epoxy 2038 with Hardener 3416 and Araldite

Fig. 7.3. The glass bead during treatment in the CR-20/CT-32 cycle.

are the two brands most commonly used in glass repair. The new "super glues" made of cyanoacrylate are used quite often to piece glass together quickly. After using the cyanoacrylate, epoxy is flowed into the cracks with an artist's brush to permanently glue the pieces. It is exceptionally difficult and time-consuming to gap-fill glass. Considerable work and experience are required. The problem of matching transparent glass colors is equally difficult. All these problems are discussed in greater detail in *Conservation of Glass*.[7]

As is the case in all conservation, the conservator must be able to recognize what the problems are and know what can be used to counter them. In glass conservation, when lead oxides are found on glass, they can be removed with 10% nitric acid; 1–5% sulfuric acid can be used to remove iron oxide, to neutralize the alkalinity of glass that is breaking down, and occasionally to remove calcareous deposits. Calcareous deposits are commonly removed with 10% hydrochloric acid and, on some occasions, by immersion in 5% EDTA, tetra sodium. Iron stains are commonly removed with 5% oxalic acid or 5% EDTA, disodium.

Realistically, few problems, other than reconstruction and restoration, are likely to be encountered on any of the glass objects found in archaeological sites dating from the mid-eighteenth century to the present. Essentially, the same chemicals and equipment required for treating ceramics are used for conserving glass.

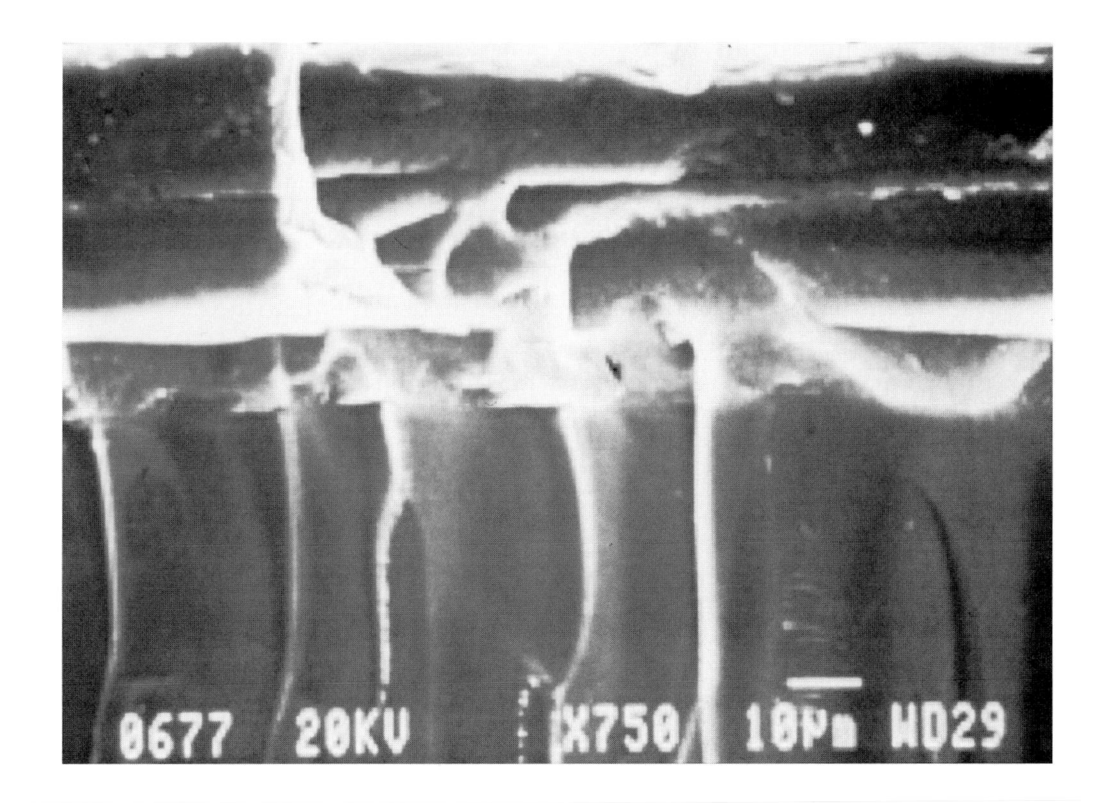

Fig. 7.4. Posttreatment cross-sectional, microscopic (X 1,200) view of the surface of the treated glass. Three distinct polymer layers are visible.

Case Study: Preservation of Seventeenth-Century Glass Using Polymers

Long-term stabilization of waterlogged glass from Port Royal, Jamaica, has been a problem. Because of the lack of controlled-environment curation, many of the bottles preserved using conventional polyvinyl acetate (PVAC) treatments have been lost due to exfoliation of surface layers and eventual breakdown of the glass. Experimentation, aimed at developing conservation strategies to preserve this assemblage, has generated new tools and questions regarding artifact conservation using polymers. Because of environmental factors, PVAC polymer processes are ineffective in stabilizing the Port Royal glass assemblage. This has fueled our experimental research into the use of siloxane polymers for archaeological preservation.

While PVAC is considered a stable consolidant that is resistant to yellowing, it has a low glass transition temperature (T_g) (28°C) and is prone to cold flow.[8] Dust and dirt tend to accumulate on the surfaces and deep crevices of glass treated with PVAC.[9] Both these factors appear to affect the Port Royal glass assemblage. Over time, bottles exposed to temperature variations, humidity, and ultraviolet light have developed thick layers of oxidation on their surfaces. In some cases, the outer surfaces of the bottles have exfoliated, resulting in the loss of diagnostic attributes.

One interesting avenue of research has been the use of aminopropyltrimethoxysilane, a common silane coupling agent, which has been used to facilitate the bonding of two pieces of glass when epoxy systems are being used as adhesives to repair broken artifacts. Coupling agents improve the adhesion of polymers to glass.[10] Because of its ability to retain strong Si-C bonds, CR-20 was included with other polymers in an experiment to determine which viscosities (centistoke) of polymers best preserve archaeological glass.

CR-20 is a trifunctional monomer resistant to solvents and photo-oxidation.[11] When used as a pretreatment to the edges of glass being repaired using an adhesive, silane groups in the monomers act to reduce water absorption into the adhesive joint, resulting in a longer-lasting bond with better adhesion qualities. This means that the junction of the repaired pieces of glass is less prone to absorption of atmospheric moisture than the glass surrounding the site of repair. The ability of silane to form strong Si-C bonds and to reduce moisture absorption makes it potentially beneficial for the stabilization of archaeological glass.

Glass made in England during the seventeenth century consisted of approximately 75% sand (silica) and a fluxing agent made from plant ash. Fluxing agents were used to reduce firing temperatures below 426°C. Glass formed at lower temperatures are susceptible to hydration. Lime was added to the silica/flux mixture to make the resulting glass harder, more durable, and resistant to hydrolysis. A two-stage reaction occurs on the surface of the glass during extended periods of immersion in water. Sodium ions are exchanged for hydrogen ions, resulting in a reduced volume of the surface glass. This exchange produces a surface layer of material that is more siliceous than the original glass. Higher Si content in the crusted layers of glass suggests that extensive ion exchange has occurred.

All the onion bottles in the Port Royal assemblage are dark grayish green. According to Plenderleith and Werner, this color is due to high concentrations of iron oxides.[12] To remain chemically stable, glass should contain at least 70% silica and not more than 16% alkali metal oxides. Additionally, glass requires between 5% to 10% calcium oxide. Glass containing higher concentrations of alkali metal oxides is potentially hygroscopic. Alkali ions will diffuse and leach throughout a porous, silica network attracting other ions and mineral-rich sediments surrounding the glass. This can potentially result in increased calcium in hydrated glass. In turn, these elevated calcium levels are associated with extensive surface crazing that reduces the transparency of the glass and results in occasional spalling of the surface. Accordingly, the outer surfaces of many bottles in the assemblage are pitted and frequently concreted with white carbonate deposits.

Since the sinking of Port Royal in 1692, the glass assemblage has been subjected to hydrolytic attack under pressure. Selective leaching of soluble materials, including alkali metal oxides, has resulted in a high silica residue in the weathered crust on the surface of the glass. Analysis of the outer crust indicates that alkaline ingredients, originally essential in the forming of the glass, are missing and have been replaced with water.

The purpose of this experiment was to evaluate the ability of selected short-chain, silanol-ended polydimethyl siloxane polymers to conserve similarly sized shards of devitrified glass. The mechanism of impregnation used to introduce these polymers into each glass shard was acetone/polymer displacement. This is a commonly applied process using the rate of evaporation of acetone, at reduced or ambient pressure, to displace acetone with a polymer solution.[13] Catalyzation of the polymer-impregnated glass was accomplished using vapor deposition of a tin-based catalyst. All treated samples were then subjected to accelerated weathering tests to determine if each polymer protected the glass from environmental damage.

Pretreatment Analysis of the Glass

Prior to treatment, scanning electron microscopic analysis and neutron activation analysis were conducted on all glass shards. Microscopy revealed a distinct outer layer or outer crust on the samples (fig. 7.5). The inner surface of the weathered crust, closely associated with the solid surface of the bottle, contained low proportions of calcium and high readings of magnesium and silica. The inner core of glass contained higher proportions of calcium. This variability resulted from absorption of magnesium from seabed sediments with consequent leaching of calcium from the core of the glass outward to the surface of the vessel. The greatest area of chemical exchange appeared to be between the inner edge of the weathering crust and the outer surface of the inner glass core. The crust's silica content was higher than that of the core glass. Distribution of elements in the crust and core glass edges are compared in table 7.1.

Side view micrographs of the surface of glass samples from Port Royal indicated that the surfaces are generally coated with a thick crust of oxidized glass, soluble and insoluble salts, and concretion from close association

Table 7.1. Sample Locations and Elements

	Element (parts per million)							
	Na	Mg	Al	Si	K	Ca	Fe	Cl
Location of Sample								
Crust								
Outer Edge	.44	3.09	4.69	69.97	5.74	9.17	2.96	2.21
Inner Edge	.41	4.88	1.96	73.50	4.07	6.11	7.13	1.07
Glass Core								
Outer Core	.65	1.33	.89	53.14	7.76	30.96	3.62	.74
Inner Core	.53	1.15	.74	52.89	8.26	31.63	3.12	.97

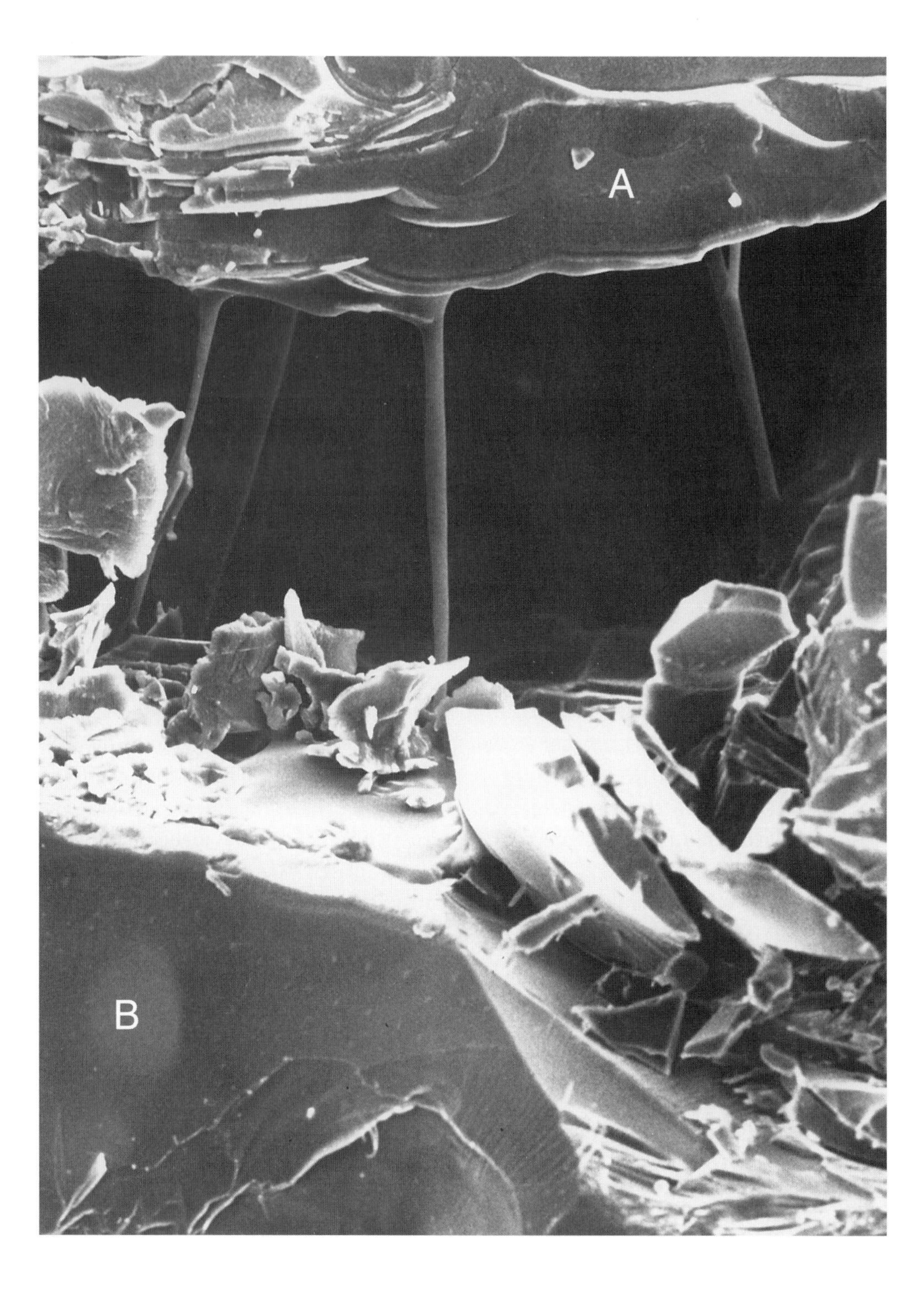

Fig. 7.5. Port Royal onion bottle OBG4: *(A)* outer surface; *(B)* inner surface of siliceous outer layer of glass. Silicone attachments spread outward from the core surface to the outer layer of glass.

with other artifacts. The surface crust is hydrophilic, containing amorphous silica compounds. The core of the glass is highly hydrated, resulting in a complex matrix that is difficult to stabilize.

The general instability of the Port Royal onion bottles may be due, in part, to the viscosity of PVAC V25 and its inability to bond effectively. Inspection of several bottles indicated the consolidant only partially penetrated the outer surfaces of the glass. Because the consolidant did not form an effective barrier on the surface of the core glass, deterioration was not halted. Additionally, the consolidant did not seem to act as an adhesive between layers of flaking glass, which may have caused the high rate of surface deterioration. Lower viscosity consolidants may penetrate throughout the matrix of an artifact but fail to act as an adhesive between the crust and core, resulting in poor consolidation of the glass.

Overview: Conservation of the Port Royal Glass Assemblage

Preservation of the Port Royal glass assemblage has been conducted using a process of slow desalination, followed by dehydration in successive baths of organic solvents, and concluding with consolidation by impregnating the glass with polyvinyl acetate. Vinyl polymers, like PVAC, have a refractive index similar to glass and generally remain stable. Accordingly, many conservators prefer to use vinyl polymers for conserving waterlogged glass. While lower molecular weight solutions of PVAC are resistant to yellowing, they are prone to cold flow.[14] This can be a significant problem for the long-term stability of the glass.

Initially, Port Royal onion bottles preserved with PVAC appeared to be stable. Unfortunately, inadequate facilities for curation have caused several of these artifacts to become unstable. For example, exposure to high humidity and fluctuating temperatures has resulted in extensive exfoliation (fig. 7.6) and discoloration of the outer surfaces of the glass. In a few cases, entire artifacts have deteriorated.

Chemical analysis of the Port Royal assemblage indicates that the surfaces of the glass samples are generally coated with a thick crust of oxidized glass, soluble and insoluble salts, and concretion from close association with other artifacts. The surface crust of the glass is hydrophilic, consisting of amorphous silica compounds, whereas the core of the glass is highly hydrated. The resulting complex matrix is difficult to stabilize. Viscous materials such as PVA V25 do not penetrate easily into the solid core of

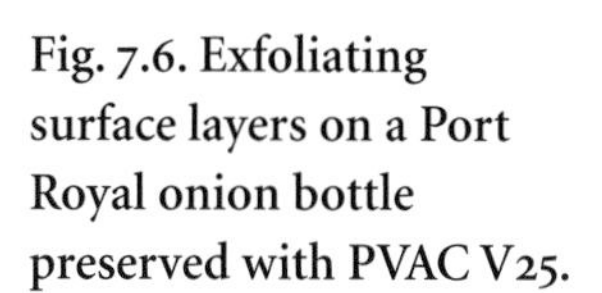

Fig. 7.6. Exfoliating surface layers on a Port Royal onion bottle preserved with PVAC V25.

glass, resulting in a superficial skin that is prone to spalling. Lower-viscosity consolidants may penetrate throughout the matrix of the glass but fail to act as an adhesive between the crust and core, resulting in poor consolidation of the glass.

In order to successfully conserve waterlogged glass with silicone oils, water within its matrix must be displaced using a volatile solvent such as acetone. Acetone is then replaced by a hydroxyl-ended, functional polymer, mixed with a cross-linking agent. To facilitate displacement of acetone with polymer, each sample was placed into a vacuum chamber. At reduced pressure, the boiling point of acetone is lowered, and thus acetone is rapidly vaporized and driven from the glass. Rapid solvent vaporization allows the polymer/cross-linker solution to permeate the matrix of the glass more easily.

Materials

The first onion bottle shard, OBG1, was treated with PR-10 polydimethyl siloxane, which is a hydroxy-terminated silicone oil with up to 5% dimethyl cyclosiloxane added. PR-10 was the longest chain silanol-ended polymer used for this experiment. It is a viscous, odorless polymer (1200 cts). OBG2 was treated with Dow Corning PA Fluid (6–10 cts), a short-chain silanol-ended, poly-

Table 7.2 Treatment Summary for Glass Fragment Samples

Sample	*Acetone Dehydration*	*Silane*
OBG1	Acetone	PR-10
OBG2	Acetone	PA fluid
OBG3	Acetone	KP80
OBG4	Acetone	4-7041
OBG5	Acetone	no silanol
OBG6	No acetone	no silanol

Note: For all samples, cross-linker was 3% CR-20; catalyst was CT-32.

dimethyl siloxane, with octamethylcyclotetrasiloxane and decamethylcyclopentasiloxane added. OBG3 was treated with KP80 (30–40 cts), which is also a silanol-ended, polydimethyl siloxane. OBG4 was treated with Dow-Corning 4-7041, a very short chain silanol-ended, polydimethyl siloxane (2–5 cts). No silicone oil was applied to OBG5. Instead, this shard was dehydrated in acetone and then impregnated with CR-20 cross-linker. The samples were catalyzed using CT-32 catalyst. Sample OBG6 was treated with CR-20 cross-linker without first being dehydrated. The sample was then catalyzed with CT-32.

CR-20 cross-linker is a hydrolyzable, multifunctional alkoxysilane, capable of tying two or more polymer chains together. CT-32 catalyst was applied in vapor form to polymerize the shards.

Experiment Procedures

Prior to treatment, six similarly sized fragments of glass from a broken onion bottle were rinsed in a series of 10 baths of deionized water to remove soluble salts. Each bath was exchanged for fresh deionized water after 24 hours. Each of five samples, OBG1, OBG2, OBG3, OBG4, and OBG5, were then placed into a beaker containing 400 ml of fresh, industrial-grade acetone. Immersed in the solvent, each sample was placed into a vacuum chamber and a reduced pressure of 5333.33 Pa (40 Torr) was applied for 6 hours at room temperature. Cessation of rapid bubbling was used as a visual indicator that free-flowing water and air had been displaced by acetone during dehydration of all the shards.

All the samples, were then placed into their respective polymer/cross-linker solutions and treated in reduced pressure environments for 6 hours. Sample OBG6 was not dehydrated prior to treatment. Alternately, OBG6 was immersed in 200 ml of CR-20 and treated in a reduced pressure environment of 5333.33 Pa (40 Torr) for 6 hours. Data for type of dehydration, silanol name, percentage of cross-linker applied by weight, and catalyst for the six glass samples are summarized in table 7.2.

After 6 hours of acetone/polymer solution displacement, each sample was placed into a plastic container with a tight-fitting lid. For this experiment, the lid and body of a ½ liter plastic container were placed in an inverted position so the lid formed the flat base of a containment chamber. A 40-g capacity aluminum dish was placed in the center of the base of the containment chamber. Fifteen grams of CT-32 catalyst were placed in this dish. A section of aluminum mesh was placed over the aluminum dish to form a platform on which a shard of glass could be positioned during catalyzation. Several paper towels were placed on the screen to absorb silicone oils flowing from the glass during the final stages of treatment. With each sample positioned in its containment chamber, the body of the container was placed over the lid and, with firm pressure, snapped into place, sealing each piece of glass within the container. A small hole was made in the upper surface of each chamber to prevent pressurization during the catalyzation process (fig. 7.7).

Positioned in their containment chambers, the six glass samples were placed into a vented warming oven and exposed to a constant temperature of 71_C for 18 hours. The samples were then removed from the oven and from their containment chambers and placed on lint-free cloths, exposed to air in a fume hood. All samples were evaluated for color, transparency, opacity, dryness, and surface scaling. After 24 hours in air, the samples were reevaluated using the same criteria.

Observations of Polymer Stabilization

Upon completion of the acetone/polymer displacement and catalyzation processes, all the glass shards were removed from their containment chambers and allowed to sit in fresh air. Two evaluations of each shard were conducted. After five minutes of exposure to air, OBG1, OBG2, OBG3, OBG4, and OBG6 were very natural in appearance. OBG1 remained transparent after 24 hours in the air. After 24 hours, no oxidation was noted on OBG2; the shard retained its coloration. OGBG3 appeared to darken slightly after treatment. While the shard was aesthetically pleasing during its first evaluation, slight scaling and oxidation were noted during the 24-hour evaluation. No changes were noted over the 24-hour period of evaluation for OBG4. The sample retained its opacity, dryness, and dark green color. OBG5 was slightly opaque when first evaluated. After 24 hours, the shard was heavily oxidized and extensive flaking was evident. OBG6 was slightly cloudy in appearance when first evaluated. After 24 hours, the sample had a slight iridescent shine on its surfaces. Data for all samples are compared in table 7.3.

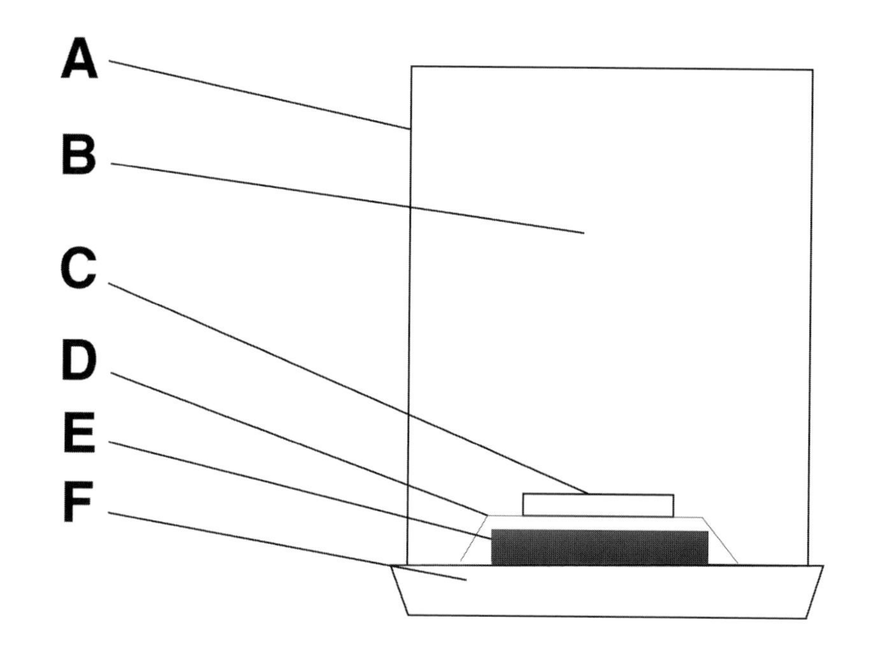

Fig. 7.7. Containment chamber configuration; *(A)* body of chamber; *(B)* warmed catalyst vapor in close proximity to artifact; *(C)* glass artifact; *(D)* aluminum screen and paper towel; *(E)* 15 g of CT-32 catalyst in aluminum dish; *(F)* lid acting as the base of the unit.

Accelerated Weathering Tests

All samples were subjected to five months of accelerated weathering testing using a Q-Panel, Q-U-V Accelerated Weathering Tester. Weathering cycles consisted of alternating 6-hour cycles of high humidity (95%) and high temperature (45°C) with a UV-A light, followed by 6 hours at lower humidity (60%) and temperature (20°C) with no light exposure. The UV-A spectrum, 315 to 400 nanometers, is the spectrum that causes the most damage for many polymers including polyethylene, PVC, acrylic, and polyvinyl acetate.

Observations—Accelerated Weathering

After weathering, the samples were taken to the Scanning Electron Microscopy Lab at Texas A&M University for analysis using electron dispersal spectroscopy and visual analysis. All samples were weighed and measured to determine what structural changes resulted from accelerated weathering. (See fig. 7.8.)

Sample OBG1 remained constant in weight and general appearance after accelerated weathering. Width measurement, taken at the widest point of the sample, indicated

Table 7.3 Observations of Glass Samples Conserved with Polymers

Sample Time in Air	*Color*	*Opacity*	*Dryness*	*Scaling*	*Appearance*
OBG1					
5 min	Light green	None	Tacky	None	Good
24 hr	Light green	None	Dry	None	Good
OBG2					
5 min	Light green	None	Dry	None	Good
24 hr	Light	None	Dry	None	Very good
OBG3					
5 min	Darker	Slight	Dry	None	Good
24 hr	Darker	Increasing	Dry	Surface flakes	Poor
OBG4					
5 min	Dark green	None	Dry	None	Excellent
24 hr	Dark green	None	Dry	None	Excellent
OBG5					
5 min	White/green	Opaque	Dry	Some noted	Poor
24 hr	White pink/green	Opaque	Dry	Extensive flaking	Very poor
OBG6					
5 min	Dark green	Slightly opaque	Dry	None	Poor
24 hr	Dark green	Slightly iridescent	Dry	None	Poor

shrinkage of 0.5%. The mass of OBG2 was reduced by 0.16% after accelerated weathering. Width measurement indicated shrinkage of 0.02%. OBG3 measured 12.60 g after accelerated weathering, indicating a reduction in mass of 0.08%. Posttreatment shrinkage in this sample was 3.68%, as measured at the widest point of the sample. OBG4 was reduced in mass by 0.19%. The width of this sample shrank by 0.12% after accelerated weathering. OBG5 was reduced in mass by 0.30% after accelerated weathering. The sample also shrank 1.01% as the result of treatment. OBG6 experienced the greatest loss of mass, with a recorded reduction of 1.03%. Physical shrinkage for the sample was 1.08%.

Aesthetically and dimensionally, OBG2 and OBG4 were the most resilient to accelerated weathering tests. During microscopic analysis, shard OBG4 was accidentally dropped on the floor and broke into two sections. The new break offered an opportunity to view the interior core of glass. A slight band of discoloration was noted. Over time, the size and appearance of this discoloration did not change

Conclusions

This experiment demonstrated a direct correlation among polymer chain length, penetration into cracks and voids within the matrix of the glass, and ability to create a protective barrier against oxidation and environmental destruction. In general, short-chain polymers appear to preserve waterlogged glass better than longer-chain polymers. This suggests that viscosity (chain length) is a factor determining the ability of polymers to displace acetone in degraded glass.

OBG1 was treated with PR-10, the longest chain polymer used in this experiment. Following processing, the shard was aesthetically pleasing. After accelerated weathering, changes in the width of the sample were noted. Results of shard OBG2, treated with PA Fluid (6–10 cts), were also very good. Evaluation of OBG3, treated with KP80 (30–40 cts), indicated that the polymer had not protected the glass from environmental destruction. A small amount of posttreatment opacity was noted on the surfaces of the shard, and after accelerated weathering, dimensional changes and flaking were noted.

OBG4, treated with 4-7041, the shortest chain polymer (2–5 cts), was evaluated as the best preserved of the glass shards. After accelerated weathering tests, color, clarity, and dimensional stability were maintained. Preservation results for OBG5 and OBG6 were predictable. The use of acetone dehydration, with no silanol to create a barrier, accelerated oxidation in shard OBG5. Since OBG6 was not dehydrated, the application of a catalyst may have bonded some of the water in the sample, preventing the high degree of oxidation and structural change noted in OBG5. Posttreatment aesthetics and dimensional changes of the six glass samples, and images of the shards are illustrated in figures 7.8 and 7.9, respectively.

Scanning electron microscopy analysis of the glass samples showed that a complex barrier coat had been formed on the surfaces of OBG1, OBG2, OBG3and OBG4. Polymers had permeated deeply into the voids of these shards, possibly forming an effective barrier within the glass. The small band of discoloration noted in OBG4 at the point of breakage was assumed to be slight oxidation resulting from exposure of interior glass to air. As noted, however, the amount and intensity of the discoloration did not change over time. Microscopic analysis of the broken edge indicated that stress fracture lines caused the discoloration thought to be oxidation. The presence of stress fracture lines also affected the coloration of glass. Fracture lines affect light refraction and glass density, resulting in a change in coloration.

Analysis of the chemical structure of OBG4 indicated that the silicone (Si) peak was substantially larger than that of calcium (Ca). In untreated control samples of glass, Si and Ca peaks were similar in proportion, suggesting that the presence of additional silicone was partially masking the presence of calcium. Iron was present in the treated glass samples. This

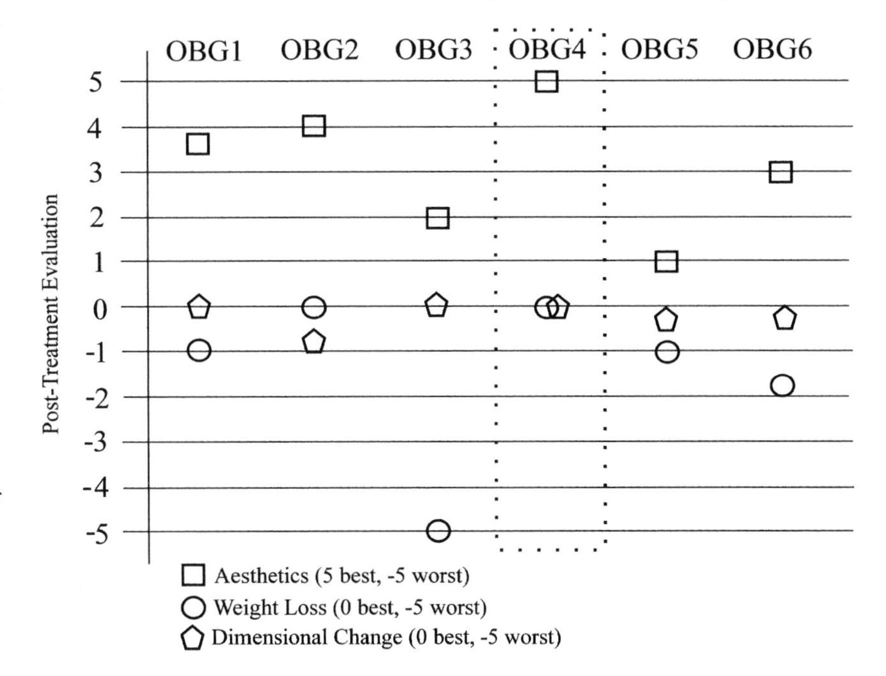

Fig. 7.8 Posttreatment evaluation of glass samples OBG1–OBG6. OBG4 was the most aesthetically pleasing glass. The shard remained dimensionally stable with comparatively little weight loss during accelerated weathering tests.

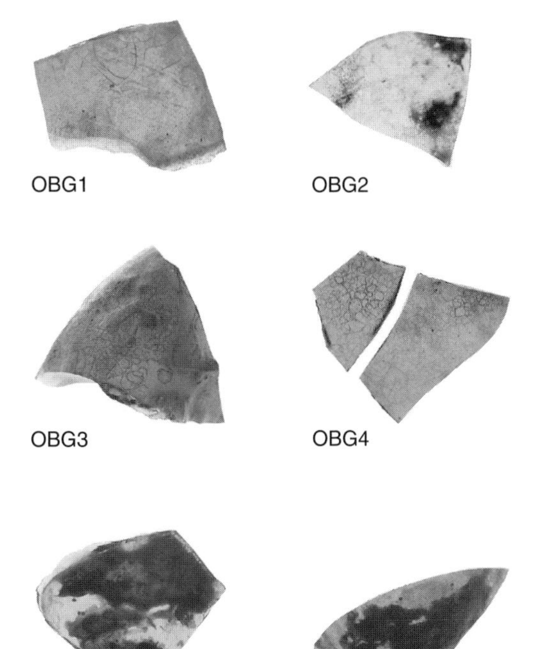

Fig. 7.9. Photos of glass samples OBG1–OBG6.

was probably due to the close association of these glass shards to iron tools and household implements within the wreckage of the submerged buildings at Port Royal.

Case Study: Preserving Waterlogged Glass and Cork

After extensive evaluation of the conservation procedures used to preserve an entire onion bottle from the excavations at Port Royal, Jamaica, the neck section of another bottle with its cork still in place was chosen for conservation by silicone bulking using PR-14 silicone, which is the most viscous polymer generally used in our laboratory. Bulking the cell structure of waterlogged corks with PR-14 silicone had been very successful, and we believed the potentially greater bulking capacity and increased cohesive strength of this silicone were advantageous in conserving this artifact. Because successful bulking and stabilization of archaeological waterlogged samples of glass and cork had been conducted, it was felt that their combination as a composite artifact posed no new problems or concerns for the stabilization of this onion bottle neck. Prior to treatment, a large section along the length of the bottle neck was noted to be loose to the touch.

Before conservation, this artifact, along with numerous other onion bottles, had been stored in a bath of fresh tap water for at least two years. The bottle neck was carefully cleaned in a fresh tub of water, using fingertips to remove any algae and debris on the surfaces of the glass. The cork appeared to be in an advanced state of disintegration. Due to its fragile state, no attempts were made to clean its deep cracks and surfaces.

Because of the artifact's small size, the dehydration process was accelerated in comparison to that used for the complete onion bottle, which required lengthy dehydration to ensure that as much water as possible was removed from the concreted sediments on the interior of the bottle. To avoid additional stress to the bottle caused by differences in temperature, the bottle neck was placed in an initial dehydration bath of acetone, with great care being taken to ensure that both the acetone and rinse water were as close to the same temperature as possible.

After 24 hours, the artifact was removed from the bath and placed into a clean beaker with new acetone. After an additional 24 hours, the artifact was removed from this bath and placed into another clean beaker with more new acetone. The beaker containing the artifact was then placed into a vacuum chamber mounted in a chest freezer. With the temperature maintained at 0°C, a vacuum of 26.5 Torr (3533.033 Pa) was applied to the artifact in its acetone bath for 8 hours. After 4 hours of applied vacuum, vigorous bubbling of the acetone had decreased substantially, and for the remaining 4 hours, smaller and less frequent bubbles were observed in the acetone solution. At this point, the vacuum was turned off and the artifact was allowed to sit in the freezer for 12 hours.

With dehydration completed, the artifact was placed into a clean beaker, and PR-14 silicone was quickly poured into the container so that the artifact was completely immersed in the silicone solution. The PR-14 siloxane had been stored alongside the sample inside the freezer to ensure that the artifact was not subjected to additional stresses caused by differential temperatures during the transfer to the bulking solution.

After carefully decanting off the last bath of acetone, we quickly transferred the artifact to a clean beaker and added PR-14 silicone until the bottle neck was completely submerged in the fluid. The beaker containing the artifact and bulking agent was then placed back into the freezer-mounted vacuum chamber. As before, a vacuum of 26.5 Torr (3533.033 Pa) was applied for 8 hours. The vacuum was then turned off and the artifact was stored overnight in the freezer.

After approximately 12 hours, methylhydrocyclosiloxane (5% by volume), a cross-linking agent, was added to the PR-14 solution and a vacuum of 26.5 Torr (3533.033 Pa) was applied for 4 hours as a means of drawing this agent into the matrix of the glass. Bulking as an additive process, in which cross-linkers are added to the bulking compound, tends to leave voids in cells and within the physical structure of the artifact more completely bulked than when cross-linkers are applied as a separate bath. In a two-stage process, the act of bulking the artifact with cross-linking chemicals appeared to remove free-running silicone from structural voids within the artifact. Because we wanted to fill the voids of the glass in the hopes of sealing it from air and other detrimental elements, the additive process seemed to be the better conservation approach. It would ensure that as much silicone as possible was left in contact within the microfissures and small voids of the glass.

With the removal of free water and displacement bulking with silicone and cross-linking agents complete, the bottle neck was removed from the solution and all excess fluids were allowed to drain off. After a couple of minutes, all free-flowing silicone had drained from the surfaces of the artifact and the bottle appeared to be thinly coated in silicone. A warming oven was preheated to 57°C. To expose the artifact to the concentrated fumes of the catalyst, a containment chamber was placed in the warming oven. Consisting of an inverted polypropylene pail with a tight-fitting lid acting as the base of the chamber, this device is inexpensive and easily adaptable for variations on the curing process.

Two ounces of CR-22 catalyst were placed on a flat dish in the center of the base of the chamber. A piece of aluminum screen was placed over the top of the dish to form a platform for the bottle neck during the curing process. A paper towel was placed on the screen to absorb any additional free-flowing silicone. The bottle neck was then placed into the chamber and allowed to sit for 4 hours (fig. 7.10). The artifact was then inspected to make sure that the excess silicone had not dripped into the catalyst. A thin layer of crusted silicone had formed on the surface of the catalyst, so the dish was removed momentarily, and two more ounces of CR-22 were added to the tray. With fresh catalyst in place, the warming oven was closed and the artifact was allowed to cure undisturbed for 36 hours.

After curing, the surfaces of the bottle and cork were dry to the touch and the paper towel, which had absorbed excess silicone, had become stiff with polymerization silicone. The bottle neck was removed from the oven and allowed to sit at room temperature for approximately 7 hours. Two days after being removed from the warming oven, a small sample of glass was removed from the base of the loose section of the bottle neck. Additionally, a small sample of cork was taken from the interior surface of the cork. These samples were set aside for electron microscopic analysis. As a final step, the fresh-break surface of the glass was sprayed with several thin coats of Krylon 1301 clear spray to ensure that the matrix of the glass was protected from the environment.

Microscopic Analysis

Scanning electron microscopy was conducted at the Electron Microscopy Center at Texas A&M University. To determine the degree of bulking that had occurred in the cork and glass samples, scanning electron microscopy was necessary to both view and then implement X-ray microanalysis as a means of identifying features in the cross sections and to determine basic chemical compositions. Using a Joel JSMT 330A scanning microscope set at a working distance of 30 mm, a series of pictures were taken of the cross-sectional views of both samples.

Inspection of the cork sample revealed that heavy bulking had occurred. In figure

7.11, the cross-sectional microscope view of the cork magnified at 1000 times, reveals that nearly all surfaces of the cell structure are heavily coated with silicone and, in many cases, the cells appear to be completely filled. SEM images of the glass neck remnant surrounding the cork indicate the formation of a complex layer of polymer throughout all the cracks and voids of the glass.

X-Ray Microanalysis

To verify the observations made during microscopy analysis, additional analysis of the chemical composition of the elements identified in the cross sections of glass and cork was necessary. Using a Joel JSM 6400 scanning microscope, X-ray microanalysis of cross section features of both samples was conducted. Prior to analysis, all samples taken from the bottle neck and cork were coated in gold/palladium and then mounted on stubs. Because this microscope was designed to allow the operator to take photographs of samples as well as give nearly instantaneous charted analysis of the elements found within features of the samples, it was the ideal tool for analyzing the cross sections of cork and glass.

The targeted area for analysis was high in silicone with an intensity reading averaging over 640 counts per second. Carbon, oxygen, magnesium, and calcium were also recorded in this sample although these elements, as well as the gold and palladium coating on the sample, were well below the intensity readings of silicone in the glass. Figure 7.5 is a 1000-times magnification of a cross-sectional view of silicone-bulked archaeological glass. Analysis of the tendrils, as indicated by the target zone labeled *A*, revealed that they were almost entirely silicone. Similar findings were noted on other areas of this sample.

Analysis of the silicone-bulked cork sample was most impressive. When viewed at 1000-times magnification, the cell structure and surface voids of the cork sample appear to have been heavily filled with silicone. Targeting the large mass in the center of figure 7.11, analysis indicated that this area of the sample was heavily bulked with silicone. Small amounts of carbon, oxygen, gold, and palladium were also noted in the target area but in much smaller amounts in comparison to silicone. Additional readings were taken along the cell walls in several areas. With a slight increase in the readings of carbon in these areas as a result of cell wall structures, silicone remained the dominant substance identified.

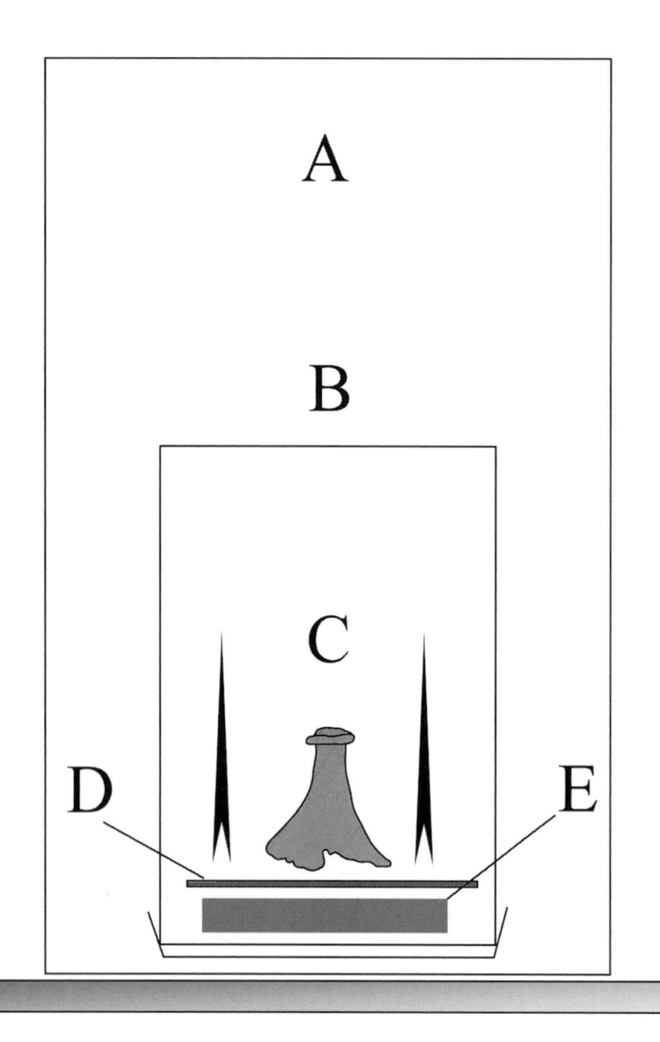

Fig. 7.10. Curing chamber and warming oven setup: *(A)* warming oven; *(B)* containment chamber; *(C)* neck of bottle with cork; *(D)* aluminum screen on which artifact sits during catalyst deposition; *(E)* aluminum tray holding 57 g of CR-22.

Observations

The assumption that the glass and cork in this waterlogged onion bottle neck could be conserved successfully as a compound artifact appears to have been correct. This may

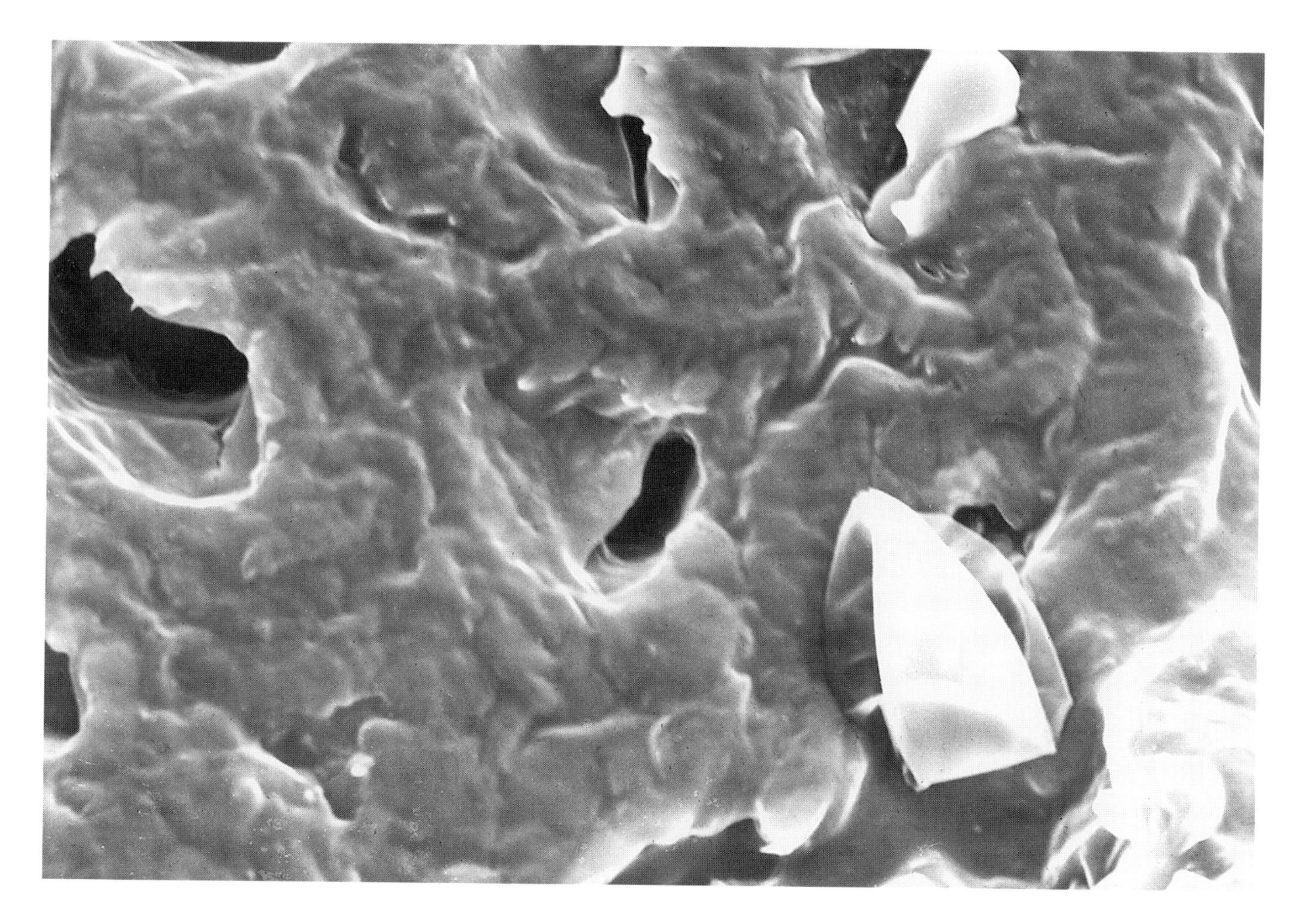

have been possible in part because of the highly degraded state of the cork. Microscopic observations and X-ray microanalysis determined that voids and cell structures of both materials were heavily bulked with silicone. The solid state of the silicone within these structures further indicates the effectiveness of the processes of cross-linking and catalyst polymerization.

The process of bulking these artifacts using an additive process in which cross-linking agents were added to PR-14 ensured the most complete bulking possible. It would be interesting to replicate this experiment using lighter molecular weight siloxanes such as PR-10. Because of their smaller size, these compounds may penetrate into the glass well but, characteristic of their size, they should not offer as much mechanical strength and cohesive bonding ability within the matrix of the devitrified glass. Unlike the end surfaces of the cork, which are visible and feel slightly rubbery to the touch, the interior and exterior surfaces of the onion bottle glass feel dry and natural in texture. In both materials, PR-14 has acted to fill cellular voids, as well as cracks and cavities, and seems to be an excellent optical bridge since it is not shiny or unnatural in appearance. Several thin coats of Krylon 1301 clear spray were applied to the outer surfaces of the bottle neck, acting as an additional sealant against environmental exposure.

Fig. 7.11. Microscopic (X 1,000) view of silicone oil–treated waterlogged cork.

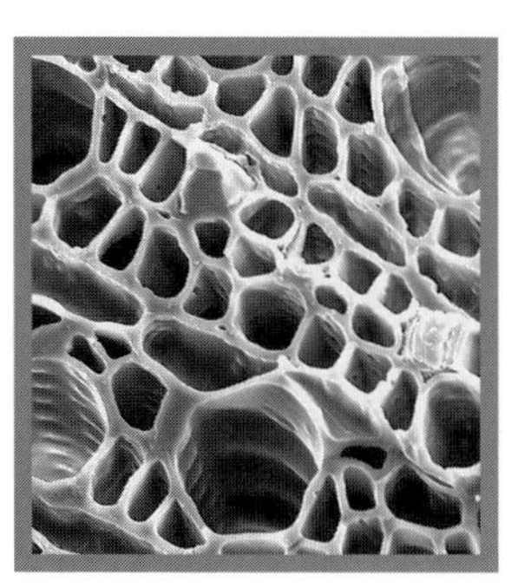

CHAPTER 8

Ivory and Bone

Ivory and bone artifacts have been recovered from nearly all excavations conducted by Texas A&M University and the Institute of Nautical Archaeology. Excavations of the seventeenth-century submerged city of Port Royal, Jamaica, which perished in a catastrophic earthquake in 1692, have uncovered bones in cooking pots from meals being prepared at the time. Skeletal remains of young children who perished when they were trapped indoors have also been recovered.

Excavations carried out by the Texas Historical Commission on the vessel *La Belle* have led to the recovery of the bones of at least two sailors. Buried in the anaerobic environment of Matagorda Bay, off the south Texas coast, these remains were so protected that a large volume of brain matter was well preserved in the cranium of one victim. Small pieces of cartilage were also recovered, and generally, the skeletal remains were in excellent condition. While skeletal materials often survive well in marine environments, bones recovered from land sites may be in poorer condition, depending on soil acidity, pH, and the presence of water.

Basic Structural Differences

The structure and function of the components of bone are complex. From an archaeological perspective, however, the major organic component of skeletal materials is ossein (protein collagen), and the main inorganic materials are calcium phosphate associated with carbonates and fluoride. Viewed under a microscope, bone has a coarse grain structure made up of collagen fibers with lacunae or voids throughout. A crystalline inorganic material called hydroxyapatite surrounds these collagen fibrils, providing strength to the bone. The chemical structure of hydroxyapatite is generally listed as $Ca_{10}(PO_4)_6(OH)_2$.

Bone and ivory are anisotropic, meaning that their collagen fibrils and associated structures have directional properties. Bone and ivory artifacts have different mechanical and strength characteristics along their lengths, widths, and thicknesses. Accordingly, artifacts made of these materials are structurally weakened when they are altered and used for manufacturing purposes. As the structural layers are removed or altered, as in the case of carvings, the support mechanism of the bone is weakened.

An important component of ivory is dentine, which is very hard and dense. When viewed through a microscope, this matrix looks very compact in cell structure with very few voids. Characteristically, ivory appears to have a network of lenticular areas, which result from the intersection of a system of striations radiating out from the center and increase in number as the tusk grows. As ivory ages or becomes waterlogged, this radiating pattern may appear similar to growth rings in a tree.

Prolonged immersion in water will cause hydrolysis of the ossein, resulting in potential warping and deterioration of an artifact. Care must be taken with heavily degraded artifacts to prevent loss of diagnostic attributes due to collapse of the cellular structure. Inorganic salts and hydroxyapatite making up the rest of the structure are affected by acidic conditions. Over time, waterlogged bone and ivory absorb organic and inorganic salts. Desalination, therefore, is an important phase in treatment of artifacts from saltwater environments. If salts are not removed, artifacts may become distorted and, in some cases, suffer from surface exfoliation as crystalline structures form during air-drying.

Bone and ivory recovered from terrestrial sites are affected by acidic soil. Hydroxyapatite dissolves in acidic conditions, leaving unsupported anisotropic collagen fibrils to shrink and warp as the artifact is air-dried. In alkaline conditions, the organic collagen hydrolyzes and is often attacked by bacteria and other microbial activity. This often results in the hydroxyapatite becoming brittle. Thus bone recovered from alkaline soils may be structurally unstable and very crumbly.

Passivation Polymers have proved to be useful for the preservation of bone and ivory from marine and terrestrial sites. Three case studies are presented in this chapter. The first is an example of consolidating friable bone from a land site. The second study illustrates a successful process for preserving delicately carved ivory from nautical excavations at Tantura Lagoon in Israel. The last case study illustrates the successful processes used to preserve large sections of elephant tusk from excavations of a Dutch East Indiaman, *Vergulde Draeck*, which sank off the coast of Western Australia in 1656.

Before discussing preservation strategies for bone, however, I examine various equipment configurations and how each may be used for the treatment of archaeological bone and ivory.

Equipment Setup for Very Fragile Bone and Ivory

In *The Conservation of Antiquities and Works of Art* Plenderleith and Werner described a simple vacuum chamber system (fig. 8.1). Instead of a vacuum pump, they suggested using a water pump system to supply slight vacuum to achieve thorough impregnation of a bulking agent into the porous structure of an artifact.[1] This system works well in situations where the use of a high efficiency vacuum pump may be too powerful and potentially hazardous for delicate artifacts. This form of water-driven vacuum treatment is slow but highly effective for the treatment of bone and ivory artifacts that are small, thin walled, or both. Many of the artifacts processed at the Archaeological Preservation Research Laboratory are more robust. These artifacts can be safely processed using a system similar to that proposed by Plenderleith and Werner but substituting a high volume, electric vacuum pump (fig. 8.2, *F*). Because of the presence of volatile vapors resulting from solvent dehydration, a double gas trap assembly *(C)* immersed in a container of dry ice or liquid nitrogen *(D)* acts to efficiently freeze gases escaping the vacuum chamber. Note also that the vacuum gauge *(E)* should be placed in line after the gas trap assembly to prevent damage from solvent vapors.

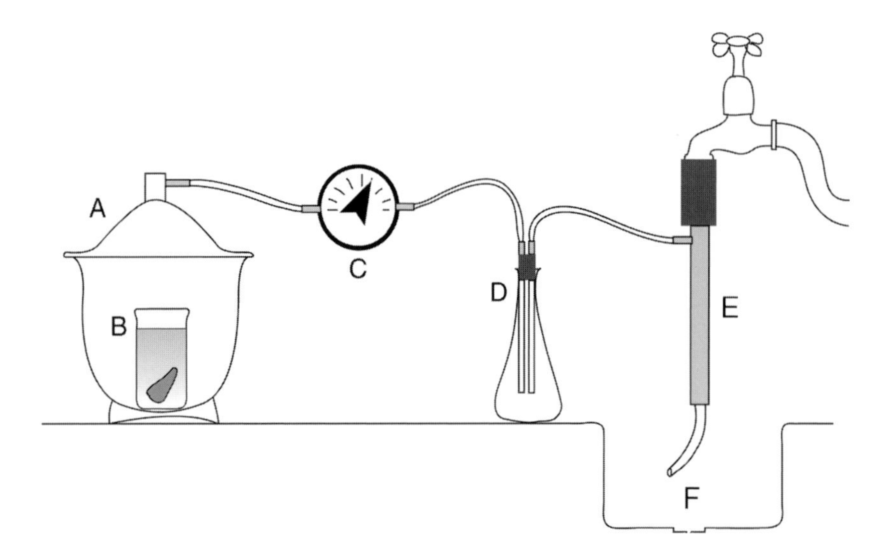

Fig. 8.1. Configuration for water flow vacuum processing of delicate artifacts: *(A)* desiccator/ vacuum jar with valve; *(B)* beaker containing artifact and either acetone or polymer solution; *(C)* in-line vacuum gauge; *(D)* in-line trap to catch solvent or polymer solution; *(E)* hose connected to running water source; *(F)* sink with drain.

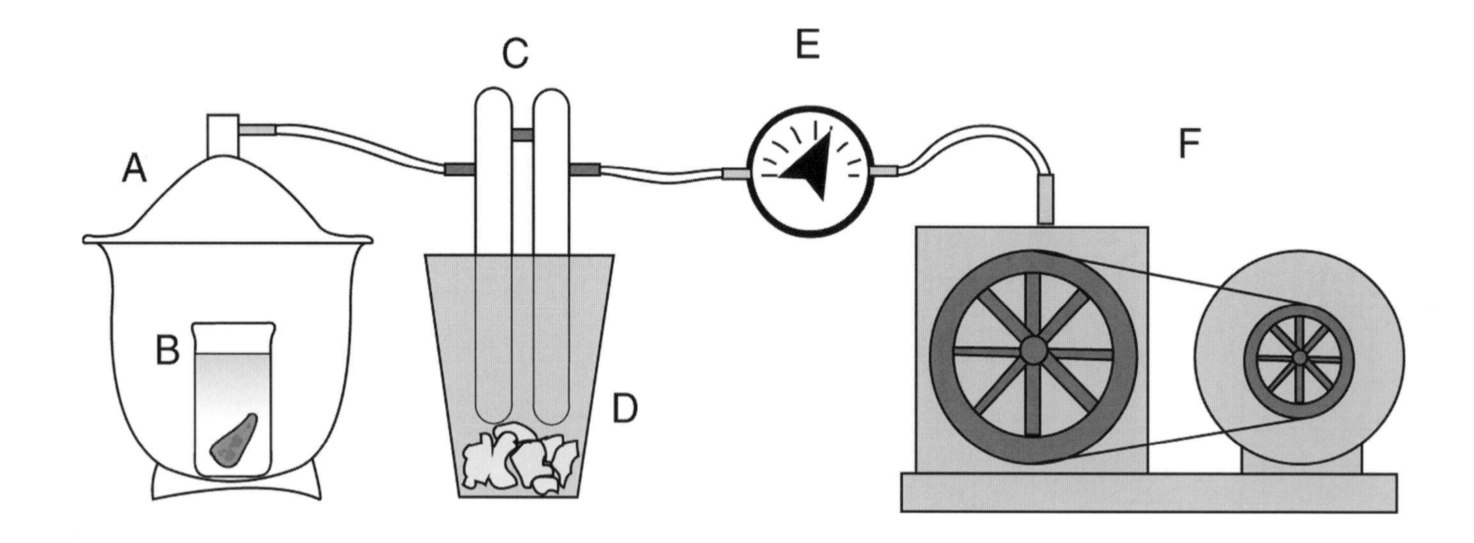

Fig. 8.2. Equipment configuration using high efficiency vacuum pump: *(A)* desiccator/vacuum jar with valve; *(B)* beaker containing artifact and either acetone or polymer solution; *(C)* in-line gas traps embedded in dry ice; *(D)* Dewar flask with dry ice; *(E)* manometer; *(F)* vacuum pump.

Case Study: Consolidating Friable Bone

Dry, desiccated bone is generally easy to consolidate using Passivation Polymers. The bone sample brought to the research laboratory had been recovered from excavations of the well-defined Clovis occupation strata at the University of Texas Gault site during a field school excavation conducted by Texas A&M University. After large clumps of soil were removed with a soft brush, the bone fragments were stored in aluminum foil. Figure 8.3 illustrates the bone's fragmented surfaces.

Before treatment, the fragmented bone, supported by the aluminum foil wrapper, was quickly rinsed in a gentle stream of tap water to remove additional dirt and dust from its surfaces. After rinsing, it was left in a fume hood for 24 hours to air-dry. The bone and its foil wrapper were then gently immersed in a beaker of acetone for approximately 1 hour to ensure that dehydration was complete. Because the bone fragments were too fragile to remove from the foil, sections of the bone were first treated with a topical application of the polymer/cross-linker solution. Because of its viscosity, CR-12, was selected for consolidation purposes. A 5% addition of CR-20 by weight was thoroughly mixed with CR-12 to create a suitable volume of polymer/cross-linker solution to consolidate the bone.

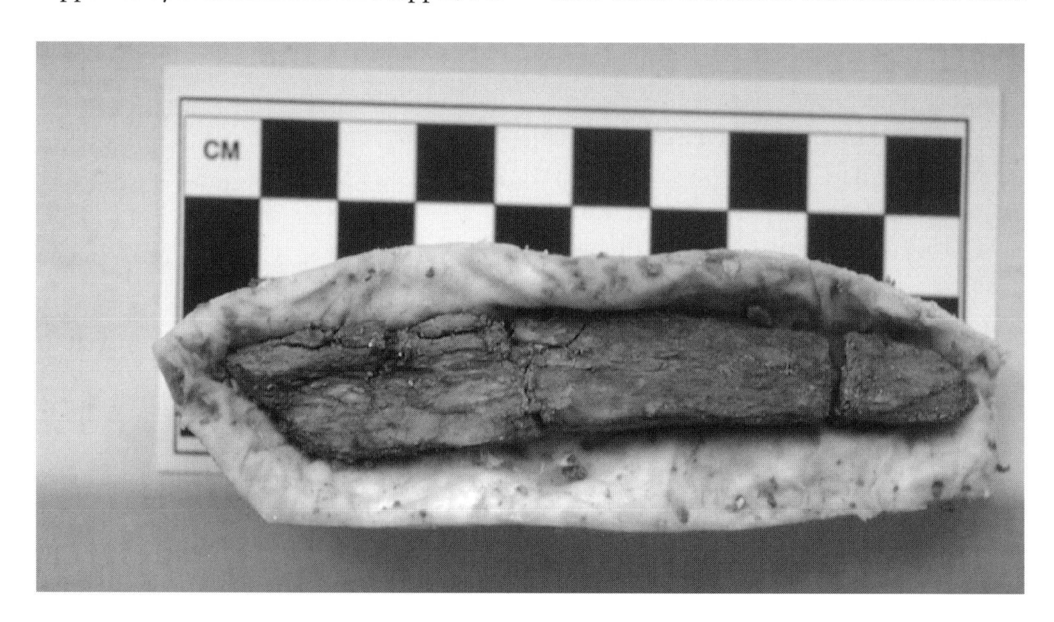

Fig. 8.3. Fragments of bone wrapped in aluminum foil.

Q-tips were used to gently apply the solution. Because the bone was thoroughly dry, the polymer solution was quickly absorbed. Repeated spot applications were made over the entire top surface until the bone was wet. The bone was then gently transferred and inverted onto a second piece of aluminum foil, and the process of applying the polymer solution was repeated.

After polymer/cross-linker solution was applied topically, the bone was more stable. The artifact was then carefully removed from the foil, immersed in the polymer/cross-linker solution, and placed into a glass desiccator/vacuum chamber, where it was treated in a reduced atmosphere environment of 28 Torr (3733.016 Pa). For a short period of time, a rapid stream of bubbles emerged from the bone. Care was taken to apply only sufficient vacuum to displace air in the bone with the polymer solution. After approximately 2 hours, only occasional bubbles were noted, and the bone was allowed to sit in the polymer solution at ambient pressure for an additional 2 hours. The fragments were then removed from the polymer solution and placed on a screen over a plastic container where runoff polymers were collected for reuse (fig. 8.4).

The bone fragments were allowed to drain of polymer solution for several hours. Soft cloths were used to blot the surfaces of the fragments to remove any remaining pooled polymer solution. The fragments were then placed into a large Ziploc bag. An aluminum dish containing 15 g of CT-32 catalyst was placed alongside the bone fragments and the bag was sealed, forming an airtight environment. Over the next 24 hours, catalyst vapors acted to polymerize the bone fragments. CT-32 catalyst has a working life of approximately 24 hours in a contained air environment. To ensure that the bone fragments were completely cured, a fresh dish of catalyst was placed in the bag and the catalyzation phase of treatment was repeated for an additional 24 hours. After 48 hours of catalyzation, the bone fragments felt dry. They were placed in a fume hood for approximately 2 hours to dissipate the smell of catalyst before analysis and labeling were completed.

Case Study: Ivory from Tantura-B Excavations in Israel

Shelley Wachsman, of the Nautical Archaeology Program at Texas A&M University, recovered a finely carved piece of ivory from a shipwreck off the coast of Israel dated to the early ninth century C.E. The artifact measured approximately three inches wide, four inches long, and one-sixteenth-inch thick. Inspection of the ivory indicated it had been intricately carved and was in very good condition. Fortunately, the artifact had been stored in seawater and transported in this state to Texas A&M University. Prior to treatment, the artifact was desalinated in a series of freshwater rinses. Initially, tap water was used. Over time, however, baths of rainwater and, finally, deionized water were used to rinse salts from the artifact. Titration tests were used to assess the rinse water. At approximately 20 ppm, titration readings were stable and the ivory was considered ready for treatment.

The artifact was placed into a beaker containing a 50:50 solution of alcohol and water

Fig. 8.4. After vacuum-assisted impregnation, the bone fragments were allowed to drain of excess polymer solution.

for initial dehydration. After 10 days, the ivory was moved to a second beaker containing fresh isopropyl alcohol, where it was dehydrated for an additional 10 days. To complete dehydration, the ivory was then transferred to a beaker containing fresh acetone. It remained in this solution for 5 days.

Because the surfaces of the artifact appeared to be very hard and smooth, there was concern that the use of a viscous polymer might inhibit thorough penetration of the polymer solution into the matrix of the artifact. Accordingly, PR-10 Passivation Polymer was selected. To a suitable volume of PR-10, a 5% (by weight) addition of CR-20 cross-linker was thoroughly mixed. The solution was then placed into a vacuum chamber and treated in a reduced atmosphere environment for approximately five minutes to de-air the solution prior to treatment. The ivory was quickly transferred to the container of the polymer/cross-linker solution. The container was placed into a desiccator/vacuum chamber, and the artifact was treated in a reduced atmosphere environment for several hours. Initially, only sufficient vacuum was applied so that a very slow stream of bubbles was noted. The chamber was locked off at reduced pressure to allow efficient acetone-silicone oil solution exchange. After approximately 3 hours, the vacuum pump was turned on, further reducing the pressure in the chamber. After 5 hours of treatment no bubbles were observed emerging from the surfaces of the ivory. The container was removed from the vacuum chamber, and the ivory was left in solution for an additional 24 hours.

After the artifact was carefully removed from the polymer solution, it was placed directly onto sheets of tissue paper. The goal was to fully support the ivory while using the paper to blot polymers from its surfaces. Sheets of tissue paper were also placed on the top surface of the artifact to remove the polymer solution. After 20 minutes, the ivory felt dry. Soft, lint-free cloths were then used to gently wipe the surfaces of the artifact, carefully removing any remaining polymer solution from carved crevices.

Catalyzation was completed using a two-stage process. First, CT-32 was topically applied to the surfaces of the artifact. Soft cloths were used to wipe off the catalyst after allowing it to sit for two minutes. The surfaces of the ivory felt smooth and had a uniform semigloss appearance. The artifact was then placed into a large Ziploc bag with an aluminum dish containing 25 g of CT-32. The bag was sealed, and the artifact was catalyzed for 24 hours. When removed from the bag, the ivory was not glossy in appearance and did not feel rubbery to the touch. Most important, no exfoliation or surface flaking had occurred. The surfaces of the artifact were lightly wiped with soft cloths, and no additional treatment was required (fig. 8.5).

Case Study: Waterlogged Tusks from Western Australia

Two sections of waterlogged and badly deteriorated tusk were shipped to the Archaeological Preservation Research Laboratory from the Western Australia Maritime Museum in Perth, Australia. These were sections taken from a cargo of long tusks that had been recovered from the shipwreck of a Dutch East Indiaman, *Vergulde Draeck*. After being recovered from the wreck, the tusks had remained in storage in freshwater. The samples sent to the lab (GD1132A and GT1373A) for treatment, however, had been wrapped in cotton gauze, which started to decay in transit. Apart from an odor problem, neither sample sustained physical damage during shipment.

Both sections of tusk were unwrapped, immediately rinsed in running tap water for 2 hours, then treated for 10 days in tap water rinse baths that were changed each day. When examined, areas of the surface of sample GT1373A were flaking. Rinsing in

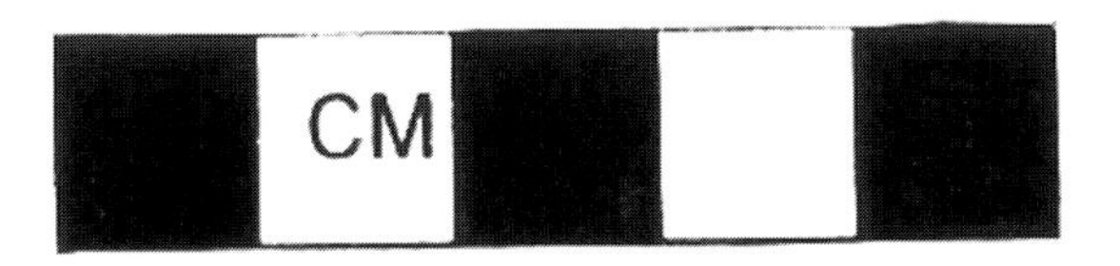

Fig. 8.5. Posttreatment view of the ivory.

freshwater did not appear to exacerbate this problem. This section of tusk measured 21.48 cm in length and 7.02 cm in diameter at its widest point. Its pretreatment wet weight was 562.2 g. Artifact GT1373A ranged in color from darker areas that registered 10Yr7/2 Munsell to lighter areas that registered 5R 2.5/2 Munsell in color. Near the base of the artifact was a roughly triangular shaped section that was bluish gray in color. This section of the tusk was slightly spongy compared to the rest of the artifact.

Prior to treatment, GT1373A was wrapped in cotton string to ensure that no delamination of the artifact occurred during treatment (fig. 8.6). The ivory was then immersed in industrial-grade acetone and placed into a vacuum chamber, where it was treated in a reduced atmosphere environment of 28 Torr (3733.016 Pa) for 24 hours. After being immersed in fresh acetone, the artifact was treated for an additional 48 hours in the same reduced atmosphere environment (fig. 8.7).

After vacuum-assisted dehydration, the artifact was transferred to a fresh beaker of acetone, where it remained for an additional 48 hours at ambient pressure and room temperature. A solution of PR-10 (40 centistoke) mixed with a 3% addition of CR-20 cross-linker (by weight) was prepared. The ivory was then quickly transferred from the acetone bath into the polymer solution. Immersed in this solution, the ivory was placed into a vacuum chamber and treated at a reduced pressure of 28 Torr (3733.016 Pa) for 96 hours. After acetone/polymer solution displacement, the artifact was removed from the polymer solution and allowed to drain of surface polymers for approximately 3 hours. The string binding was then removed, and

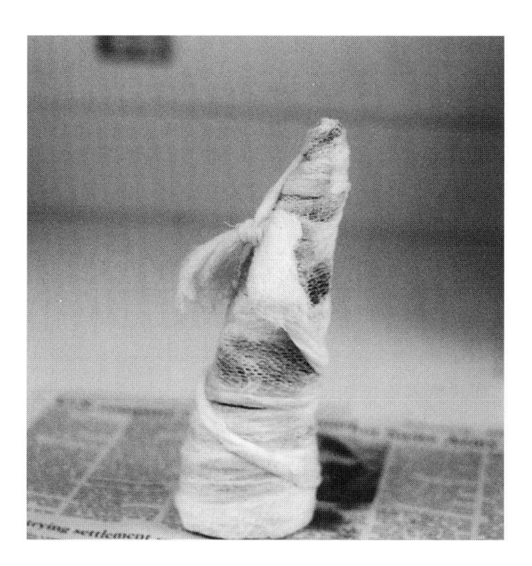

Fig. 8.6. Tusk (artifact GT1373A) bound in cotton string, prior to catalyzation.

Fig. 8.7 Tusk (artifact GT1373A) during acetone/polymer solution displacement.

the artifact was immersed in a bath of CR-20 to assist in washing additional polymer solution from surface cracks and crevices. After immersion cleaning for less than one minute, the tusk was removed from the solution and gently brushed with a soft toothbrush to help clean crevices on the surface (fig. 8.8). The tusk was then gently wiped with lint-free cloths, rebound with cotton string, and placed into a large plastic container, along with an aluminum dish containing 25 g of CT-32 catalyst. After 24 hours of exposure to catalyst vapors at 22°C, the tusk was removed and inspected.

Slight seepage of polymer solution was noted on the base of the tusk and in the area of the softer, blue gray ivory. These areas were first wiped with a cloth dampened with a few drops of CR-20 cross-linker, which acted to remove surface-pooled polymer solution. CT-32 was applied topically to the base and discolored areas of the tusk, and left on for four minutes; soft cloths were then used to wipe the areas dry of remaining catalyst. The artifact was placed back into a plastic container, where it was catalyzed for an additional 4 days, exposed to CT-32 catalyst vapors. Each day, an aluminum dish containing 25 g of fresh CT-32 was placed in the container to ensure maximum catalyzation.

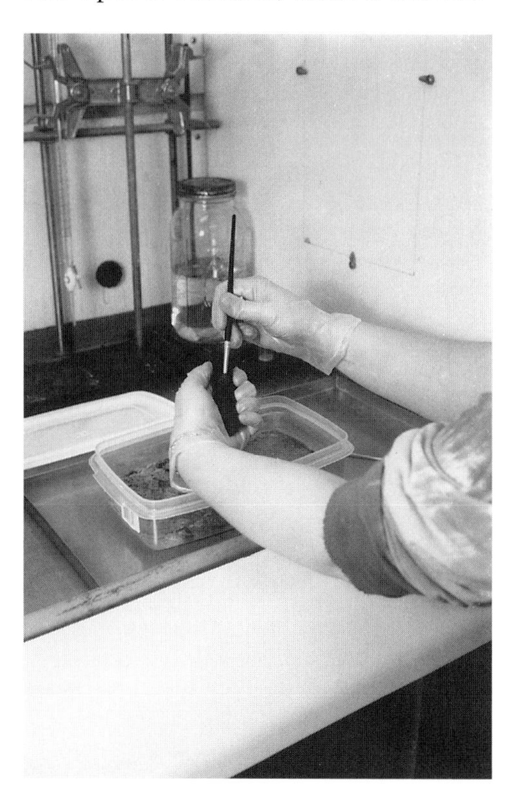

Fig. 8.8. Surface-cleaning the tusk (artifact GT1373A) with CR-20 and a soft toothbrush.

Fig. 8.9. End view of the tusks. The slightly blue gray discoloration of artifact GT1373A *(left)* is visible on the right side of the artifact near its base. Artifact GD1132A *(right)* is lighter in color. Note that the slight amount of delamination visible on the base of both tusks has not increased as the result of treatment.

The second section of tusk, GD1132A, was conserved using the same procedures. Unlike artifact GT1373A, however, this artifact did not have any blue gray discoloration. After treatment, both artifacts were photographed, weighed, and measured. In neither case was shrinkage noted, and no cracks developed as the result of treatment. To some degree, the polymer solution acted as an optical bridge, eliminating the patch of surface flaking apparent in artifact GT1373A before treatment (fig. 8.9).

Both sections of tusk have remained dimensionally stable after treatment. No seepage of polymer solution has been noted, and both artifacts remain stable in coloration. Preservation of these artifacts demanded a quick and imaginative response in trying to account for possible delamination of the tusks during treatment. Because the tusks were badly degraded, we felt it wise to tightly bind the artifacts with cotton string to prevent further delamination during treatment.

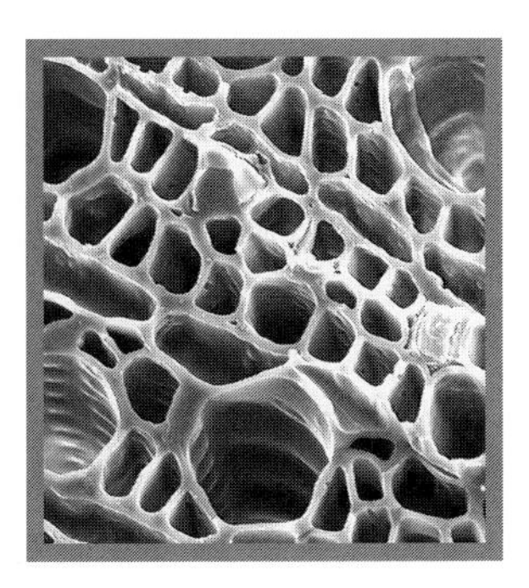

CHAPTER 9

Expanding the Conservation Tool Kit

All case studies and research discussed to this point have been directed at the preservation of organic substances by displacing water and salts with polymer and resins that, once polymerized, ensure the preservation of diagnostic attributes of organic artifacts. Silicone oil technologies are useful for other aspects of artifact preservation as well. Combining computerized tomography (CT) data and computer-aided design (CAD) modeling software, technicians can create three-dimensional models of artifacts using laser driven polymerization of each "slice" of CT data. The process permits nonintrusive investigation and recording of complex artifacts.

CT and CAD technologies can be especially useful in the preservation of iron objects from marine environments. When exposed to water, the natural tendency of iron is to revert to oxides. In a marine environment, oxides, sediments, and salts combine to form a hard outer surface as deterioration of the metal progresses. The buildup of concretion also acts to preserve the original form of the artifact. In many cases, little to no iron will remain while wood closely associated with the artifact may be preserved in perfect, waterlogged condition.

Traditionally, conservators use radiographs to look inside the concreted mass. In doing so, they can devise a conservation strategy for the artifact, based on the determination of what iron remains and how clean the cavity of the concretion is. But devising appropriate conservation strategies using traditional radiographs is difficult when dealing with large, complex artifacts. More important, it is extremely difficult to clean the cavities of intricate artifacts sufficiently to allow the flow of resins used in the mold-making phase of conservation.

Iron artifacts recovered from marine environments fall into three basic categories. The first group is concreted artifacts in which no metal remains. The second group comprises solid metal artifacts deeply imbedded in concretion due either to partial surface disintegration or to close proximity with other artifacts that have deteriorated. The last group, consisting of concreted metal objects in which only some of the metal remains, are the most challenging artifacts. In all three cases, the conservation process is often hindered by the inability of traditional radiographs to supply sufficient data to give the conservator a complete view of the artifact concealed within the concretion. Delicate details of a three-dimensional artifact are also often impossible to distinguish in two-dimensional film prints. During casting and removal of the outer layer of concretion with pneumatic chisels, fine details of an artifact may be missed unless the conservator can distinguish surface features and act quickly enough to prevent destruction and loss of internal features.

The advent of stereolithographic (STL),

rapid-prototype technologies represents a new set of tools for better diagnosis and modeling of artifacts prior to commencing conservation. This technology affords many benefits for the archaeological conservator. Three-dimensional imaging of a complex artifact allows the conservator to view the artifact from an infinite variety of angles and perspectives. Through the use of three-dimensional CAD programs, a radiographic view of an artifact can be manipulated and rotated until the conservator has the information necessary to commence conservation. To attain similar views of the interior of a concretion with conventional methods is time-consuming and often less accurate, based on the limitations of the diagnostic machinery. CT-STL analysis of an artifact allows the conservator to create an exact replica of the artifact in resin prior to treatment of the original artifact. A three-dimensional model is an invaluable aid to anyone who must clean encrustation from delicate features of an artifact. Additionally, the ability to create a resin cast replica of the artifact being treated is an insurance policy. If something goes wrong during treatment, the presence of the replica may serve to retrieve data that might otherwise be lost.

Computerized Tomography and the Stereolithographic Process

Stereolithography is literally three-dimensional printing. I have not attempted to explain the stereolithographic process and associated equipment, because for archaeological purposes, understanding potential applications for the process is far more important than understanding the mechanics of the technology. Essentially, three main components were brought together to do reproductive modeling of encrusted artifacts: a local hospital donated time and personnel to obtain computerized tomographic (CT) images of what is inside an encrustation; we identified heavily concreted artifacts that could be easily handled and x-rayed; and we found a company willing to work with us to create rapid prototype reproductions of the artifact.

CT technology is a nondestructive means of looking inside a concreted artifact. The machinery, working with high energy X rays, produces two-dimensional slices of the cross section of an artifact. Each slice is between 0.004 inches and 0.006 inches, depending on the resolution needed. Thinner slices produce higher resolution images. Naturally, producing a complete image requires many more slices when done in high resolution. Scanning time is therefore longer and all related costs are higher. Creating a CT image requires compiling a series of slices, which when complete allow the operator and conservator to evaluate the integrity of the artifact inside the concretion. Each slice is a discrete set of data about one particular section of the artifact. When the CT scanning process is complete, data that have been stored on disk during the scanning must then be processed using special software that renders the slices, creating a three-dimensional image. Programs such as MIMICS (Materialise's Interactive Medical Image Control System) are widely used for the purpose of aligning and correcting individual scanned images. Once scanned images are prepared, additional software is required to transform the data into an integrated image that is usable on rapid prototyping instruments.

The stereolithographic machinery is unique. It consists of a large vat of photocurable resin, mirrors, a UV laser assembly, and an elevator assembly that is precisely driven through the vat (fig. 9.1). As image data are processed, the laser assembly scans the layer of the part being modeled. As one layer is completed, the elevator assembly drops slightly to allow the laser to draw the next section of artifact information onto the first layer (fig. 9.2). The process continues until all the layers have been re-created. The cast is put into an oven for curing of the resins before touch-up work can be completed. Depending on the size of the artifact, the rapid prototyping system can take several hours.

Case Study: Scanning an Encrusted Artifact—CT Scanning Used as a Diagnostic Tool

A small artifact, thought to be a heavily concreted sword hilt, was chosen for demonstration purposes. The artifact had been recovered from a nautical excavation site and had been kept wet prior to treatment. This was an important step to take for two reasons. In a wet state, contrasted densities of the various materials in the concretion are accentuated, resulting in a better artifact image. Second, as a concretion dries out, it becomes harder. This makes the task of removing it more difficult, especially when it is in close association to small and delicate artifacts.

Prior to CT imaging, the artifact was visually inspected for identifiable surface features (fig. 9.3). The artifact was then mounted into the CT unit, using the headrest assembly for placement (fig. 9.4). Throughout the scanning process, individual scans were visible on the computer monitor. Naturally, each scanned image provided only a limited view of the artifact (fig. 9.5). After compiling in graphic software, however, an excellent three-dimensional image of a sword handle was created (fig. 9.6).

New Tools—New Directions in Research

The realm of digital media offers exciting application opportunities for archaeological investigations. At a time when museums are adding interactive displays and mobile exhibits, rapid prototype technology has infinite uses in the museum and research environment.

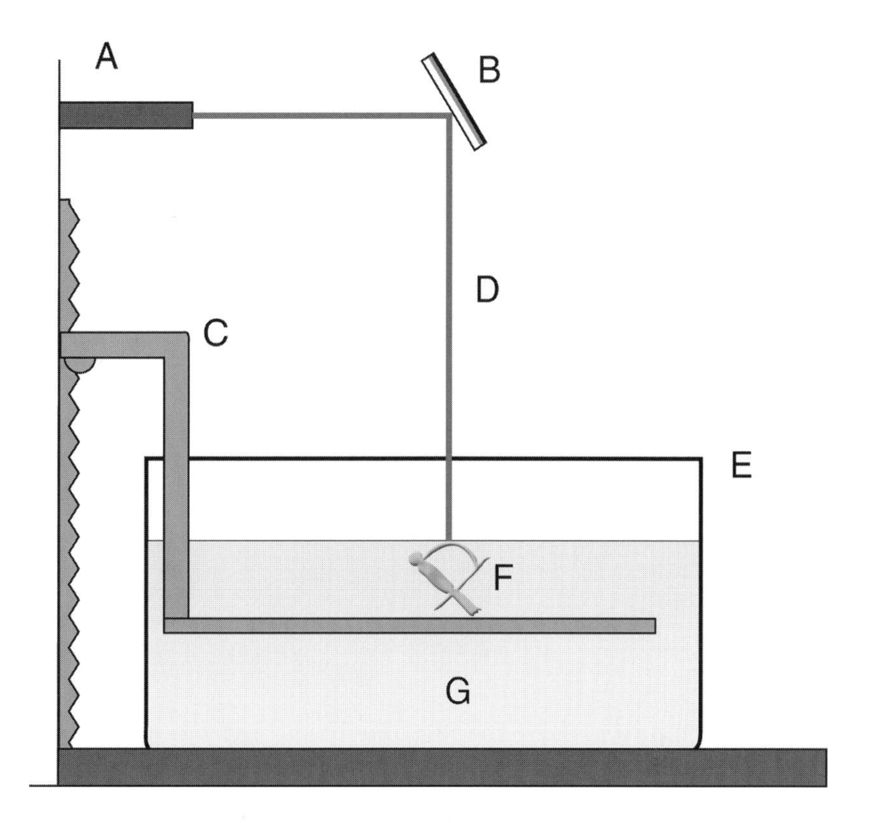

Fig. 9.1. A schematic drawing illustrating the basic stereolithographic process: *(A)* laser gun; *(B)* mirror; *(C)* elevator assembly; *(D)* reflected laser beam; *(E)* vat; *(F)* replicated artifact; *(G)* photocurable resin.

Three-dimensional images are infinitely scalable. An example will serve to illustrate the importance of scalable models for filling in the gaps in special collections. The nature of the fossil record is such that some specimens are rare. Other specimens may be more plentiful, but incomplete. Stereolithographic rapid prototyping can be used to fill in the gaps in museum collections. If, for example, the specimen in one museum is missing a single vertebra or series of vertebrae, appropriately fashioned replica can be made using samples from another museum or a more intact specimen. Even if the fossil animals

Fig. 9.2. View of the process of stereolithography as a section of resin is solidified using a laser.

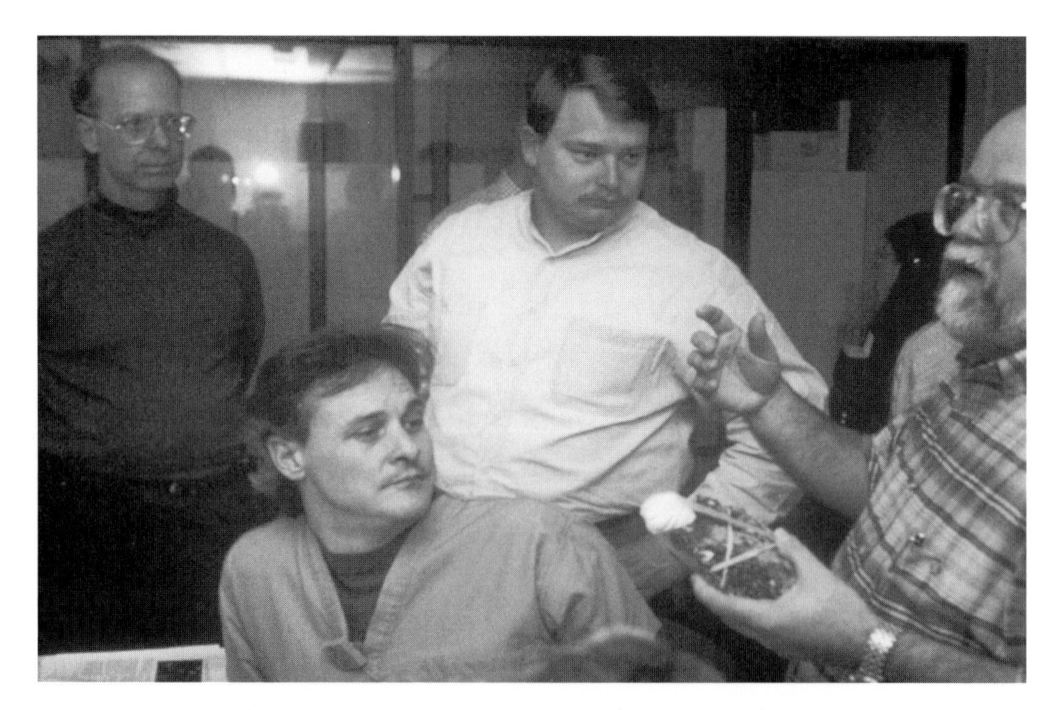

Fig. 9.3. A small concretion is discussed prior to CT scanning. Note that the encrustation is supported by rubber bands and tape.

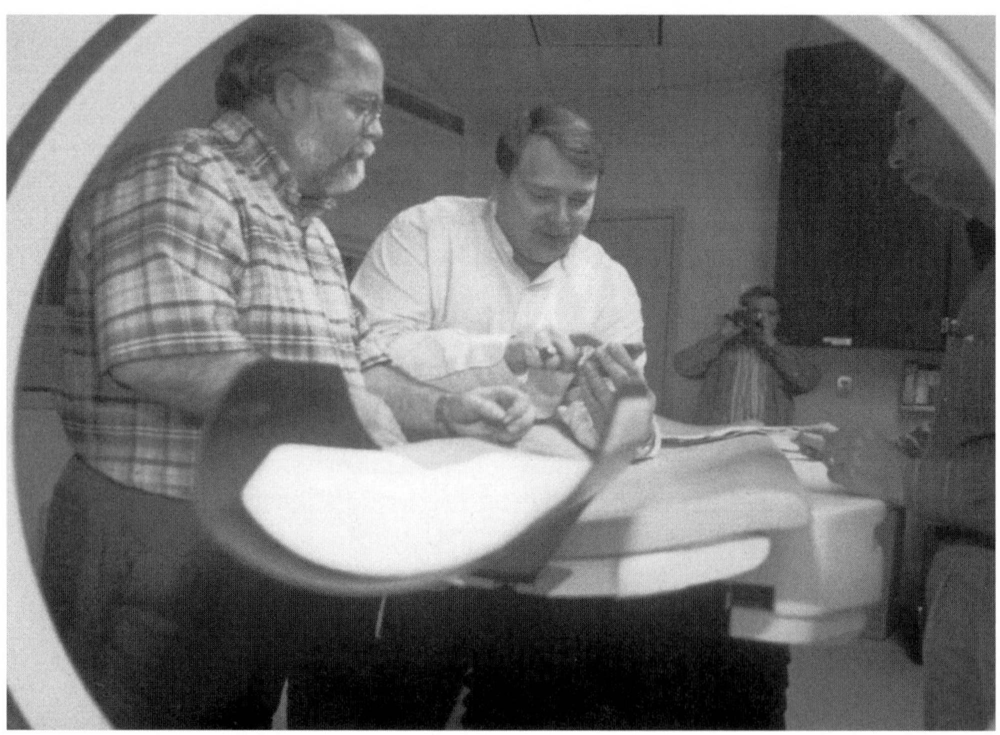

Fig. 9.4. The concretion is placed in the headrest of the CT machinery.

differ in size, appropriately sized vertebra can be manufactured to make the fossil remains complete for display purposes. Because the fabricated vertebrae are made of resin, they can be dissimilarly colored to illustrate that they are replacement parts or, when desired, colored to look like real bone.

Stereolithographic prototyping can also be used to fill in gaps of an artifact assemblage for display purposes. If for instance, a particular Clovis point is missing in an assemblage, a replica artifact can be made from a borrowed artifact. Depending on the resolution of the original CT scans, reproductions

Fig. 9.5. As individual scans are completed, they are visible on the computer screen.

can be extremely accurate in size and surface detail. Rapid prototype models can also be invaluable in the classroom. Instead of working with rare and often fragile specimens, instructors can utilize resilient resin copies that can withstand years of student handling.

Images of artifacts compiled from CT data can be rendered into three-dimensional images in CAD software. These images, which are created as part of the stereolithographic process, are quite useful for display purposes. These three-dimensional models can also be used as the basic figures to create morphed animation sequences for displays. Numerous inexpensive software programs can be used to turn start and end point images into animated sequences.

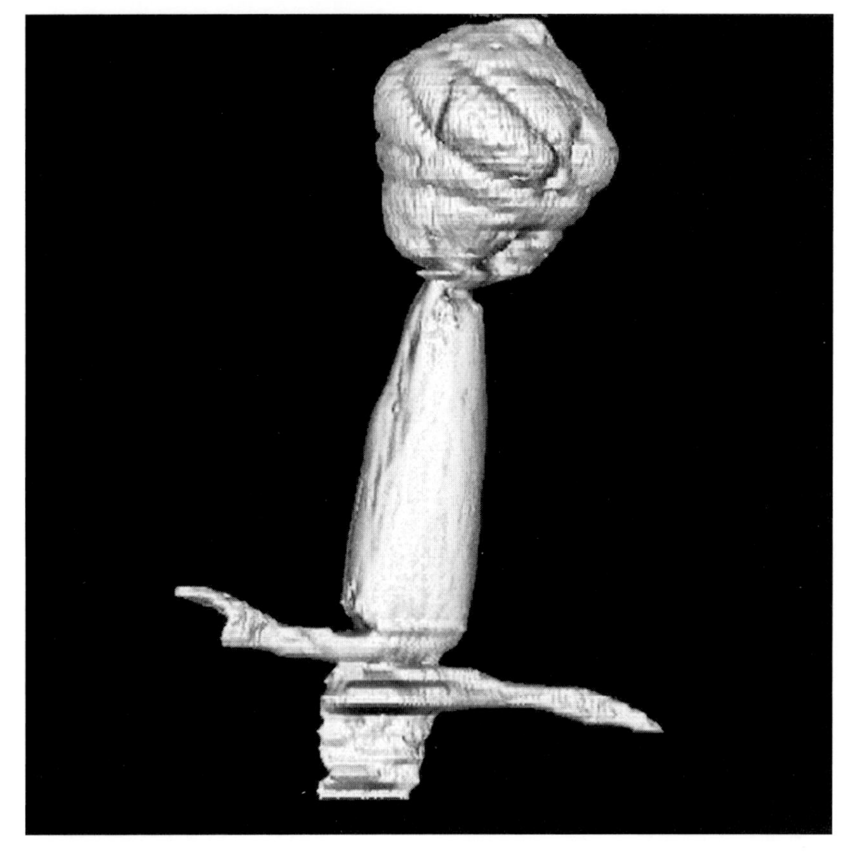

Fig. 9.6. After scanning, the individual slices of the image are compiled to create a three-dimensional image of the artifact inside the concretion.

Notes

Chapter 1

1. CR-20 is also known as MTMS (methyltrimethoxysilane); its chemical formula is $(CH_3O)_3$ $SiCH_3$. For consistency I have used the term CR-20 throughout this book.
2. CT-32 is also known as DBTDA (dibutyltin diacetate); its chemical formulation is $C_{12}H_{24}O_4Sn$. For consistency I have used the term CT-32 throughout this book.

Chapter 3

1. Charles D. Moore, "Reassembly of a Sixteenth-Century Basque Chalupa," *Material History Review* 48 (fall 1998).
2. Martha Carrier, personal communication, June 5, 2000, regarding preservation of the timbers of the Red Bay Chalupa.
3. R. H. Atalla and D. L. Vanderhart, "Native Cellulose: A Composite of Two Distinct Crystalline Forms," *Science* 223 (1984): 283; Joseph B. Lambert, Catherine E. Shawl, and Jamie A. Sterns, "Nuclear Magnetic Resonance in Archaeology," *Chemical Society Review* 29 (2000): 175–82; Michael A. Wilson, Ian M. Godfrey, John V. Hanna, Robinson A. Quezada, and Kim S. Finnie, "The Degradation of Wood in Old Indian Ocean Shipwrecks," *Geochem* 20:5 (1993): 599–610.
4. Michael A. Wilson, Ian M. Godfrey, John V. Hanna, Robinson A. Quezada, and Kim S. Finnie, "The Degradation of Wood in Old Indian Ocean Shipwrecks," *Geochem* 20:5 (1993): 599–610.
5. Allen Brownstein, "The Chemistry of Polyethylene Glycol," in *Proceedings of the IOCM Waterlogged Working Group Conference,* ed. D. Grattan (Ottawa: Canadian Conservation Institute, 1982).
6. E. E. Astrup, "A Medieval Log House in Oslo–Conservation of Waterlogged Softwoods with Polymer Glycol (PEG)," *Proceedings of the Fifth ICOM Group on Wet Organic Archaeological Materials Conference,* 1993.
7. Malcolm Bilz, Lesley Dean, David W. Grattan, J. Clifford McCawley, and Lesley McMillen, "A Study of the Thermal Breakdown of Polyethylene Glycol," *Proceedings of the Fifth ICOM Group on Wet Organic Archaeological Materials Conference,* 1993.
8. Michael A. Wilson, Ian M. Godfrey, John V. Hanna, Robinson A. Quezada, and Kim S. Finnie, "The Degradation of Wood in Old Indian Ocean Shipwrecks," *Geochem* 20:5 (1993): 599–610.
9. Ibid.
10. C. V. Horie, *Materials for Conservation: Organic Consolidants, Adhesives, and Coatings* (Oxford: Butterworths, 1987), 160.
11. David Gilroy and Ian Godfrey, eds., *A Practical Guide to the Conservation and Care of Collections* (Perth, Australia: Western Australia Maritime Museum, 1998), 91.

Chapter 4

1. J. M. Cronyn, *The Elements of Archaeological Conservation* (New York: Routledge, 1992), 274.
2. Ibid.
3. J. M. Cronyn, *The Elements of Archaeological Conservation* (New York: Routledge, 1990), 176.

Chapter 5

1. J. P. Maish, "Silicone Rubber Staining of Terracotta Surfaces," *Studies in Conservation* 39 (1994): 250–56.

2. Stefano Pulga, "A Note on the Use of Silicone Rubber Facings in the Reassembly of Archaeological Painted Plasters," *Studies in Conservation* 42 (1997): 38–42.
3. Elizabeth S. Goins, "The Acid/Base Characteristics of Sandstone, Limestone, and Marble, and Its Effects upon the Polymerization of Tetraethoxysilane," paper from the Institute of Archaeology, 5 (London, 1994): 19–28.
4. Mason Miller, personal communication, 1999. Miller, a graduate student at Texas A&M, directed studies in the university's Archeological Preservation Research Laboratory.

Chapter 6

1. Inger Bojesen Koefoed, Ion Meyer, Poul Jensen, Kristiane Straetkvern, *Report 4.1: Conservation of Wet Archaeological Rope* (Brede, Denmark: National Museum of Denmark, Section for Organic Materials, 1996).
2. Ibid.
3. Corcoran Laboratories, 5558 Springknoll Lane, Bay City, Mich. 48706, 517-892-6580.
4. Vera de la Cruz Baltizar, *Plastination as a Consolidation Technique for Archaeological Bone, Waterlogged Leather, and Waterlogged Wood.* Diploma thesis, Department of Art, Queens University, Kingston, Ontario, 1996, 110.
5. Ibid.

Chapter 7

1. Robert H. Brill, "Ancient Glass," *Scientific American* (November 1963).
2. J. M. Cronyn, *The Elements of Archaeological Conservation* (New York: Routledge, 1992).
3. Robert H. Brill, "Ancient Glass," *Scientific American* (November 1963).
4. Colin Pearson, ed., *Conservation of Marine Archaeological Objects* (Boston: Butterworths, 1987).
5. Roy Newton and Sandra Davison, *Conservation of Glass* (London and Boston: Butterworths, 1989), 179–80.
6. H. J. Plenderleith and A. E. A. Werner, *The Conservation of Antiquities and Works of Art: Treatment, Repair, and Restoration,* 2d ed. (Oxford: Oxford University Press, 1976).
7. Roy Newton and Sandra Davison, *Conservation of Glass* (London and Boston: Butterworths, 1989), 179–80.
8. Ibid.
9. G. Torrraca, *Synthetic Materials Used in the Conservation of Cultural Property* (Rome: International Centre for the Study of the Preservation and the Restoration of Cultural Property, 1963), 303–8.
10. Roy Newton and Sandra Davison, *Conservation of Glass* (London and Boston: Butterworths, 1989), 179–80; J. M. Bettembourg, "Protection des Verresde Vitraux Contre les Agents Atmosphériques, Etudes des Filmes de Résines Synthetétiques," *Verres et Réfract* 3 (1976): 87–91; R. F. Erret, M. Flynn, and R. Brill, "The Use of Silanes on Glass, Adhesives, and Consolidants," *Proceedings of the Tenth IIC Congress* (Paris, 1984), 185–90.
11. C. V. Horie, *Materials for Conservation: Organic Consolidants, Adhesives, and Coatings* (Oxford: Butterworths, 1987), 156.
12. H. J. Plenderleith and A. E. A. Werner. *The Conservation of Antiquities and Works of Art: Treatment, Repair, and Restoration,* 2d ed. (Oxford: Oxford University Press, 1976.)
13. Jerome Melvin Klosowski, C. Wayne Smith, and Donny Leon Hamilton, *Conservation of Organic and Inorganic Materials,* U.S. Patent and Trademark Office, Washington, D.C., August 4, 1998, patent 6,022,589.
14. G. Gueskens, M. Borsu, and C. David, "Photosynthesis and Radiolysis of Polyvinylacetate—3. Effects of Temperature on the Photolysis," *European Polymer Journal* (Oxford: Pergamon Press, 1972), 8.

Chapter 8

1. H. J. Plenderleith and A. E. A. Werner, *The Conservation of Antiquities and Works of Art: Treatment, Repair, and Restoration,* 2d ed. (Oxford: Oxford University Press, 1976).

Index

Italic *t* indicates a table; *f* indicates a figure.

ISBN 1-58544-218-6
90000
9 781585 442188